Stories of Modern Technology Failures and Cognitive Engineering Successes

Stories of Modern Technology Failures and Cognitive Engineering Successes

Nancy J. Cooke
Frank Durso

CRC Press
Taylor & Francis Group
Boca Raton London New York

CRC Press is an imprint of the
Taylor & Francis Group, an informa business

CRC Press
Taylor & Francis Group
6000 Broken Sound Parkway NW, Suite 300
Boca Raton, FL 33487-2742

CRC Press is an imprint of Taylor & Francis Group, an Informa business

Printed in the United States of America on acid-free paper
10 9 8 7 6 5 4 3 2 1

International Standard Book Number-13: 978-0-8058-5671-2 (Softcover)

Library of Congress Cataloging-in-Publication Data

Cooke, Nancy J.
Stories of modern technology failures and cognitive engineering successes / Cooke, Nancy J., Durso, Frank.
p. cm.
Includes bibliographical references and index.
ISBN 978-0-8058-5671-2 (hardback : alk. paper)
1. Machinery--Monitoring--Congresses. 2. Machinery--Maintenance and repair--Congresses. I. Durso, Frank. II. Title.

TJ153.C638 2007
620.8'2--dc22 2007023260

Visit the Taylor & Francis Web site at
http://www.taylorandfrancis.com

and the CRC Press Web site at
http://www.crcpress.com

Contents

Preface

This book will take you "behind the scenes" of the aftermath of problems with human–technical systems to reveal the significant efforts and the thought processes of dedicated scientists and engineers who have made a positive difference. The stories we tell begin with problems, sometimes of disastrous and tragic proportions, but our stories are not about the disasters. They are about the solution to the problem, the repair of a system that was not human- or work-centered, or the design of a fail-safe procedure to mitigate further disasters. They are not just about any solution, but specific solutions that involve cognitive systems engineering.

Chapter 1 supplies additional background, gives our perspective, and supplies details about what makes an effort like the ones reported in this book truly successful. We then present seven stories about times when modern technology failed and about how cognitive engineers were able to identify the problem and fix it. We hope that the narratives allow the reader to understand some of the issues faced by these researchers and some of the thinking they went through to reach a successful conclusion. The seven stories are brought together in two concluding chapters. In the penultimate chapter we hear directly from the heroes of our stories as they supply answers to questions about doing cognitive engineering, becoming a cognitive engineer, and the role of cognitive engineering in society. Finally, we were fortunate to have William Howell, renowned scientist and leader in human factors with significant experience at outreach, deliver an insightful commentary that addresses the human factors community and the ins and outs of marketing work in the field. Dr. Howell offers much sage advice to the professional.

Although this book targets a broad audience—virtually anyone who may be interested in learning a bit more about the new and burgeoning field of cognitive engineering—we also anticipate that it may be useful as a supplement to undergraduate and graduate courses on human factors, applied psychology, and industrial and systems engineering. The instructor of such courses might find the summaries of lessons learned and the suggested readings at the end of each story particularly helpful. In addition, we attempted to represent a variety of cognitive engineering applications, content areas, methods, and concepts across the seven stories. Chapters 2, 3, 4, and 8 cover training applications, and chapters 5, 6, 7, and 8 cover applications in design.

Military applications are addressed in chapters 2 and 8, transportation applications in chapters 3 and 4, and communications, medicine, and emergency response in the nuclear industry being covered in chapters 5, 6, and 7, respectively. Several of the chapters address highly complex systems or systems of systems (chaps. 7 and 8), whereas others touch on more traditional human–machine interfaces (e.g., chap. 6). One of the pedagogical benefits of focusing on cognitive engineering solutions is that this perspective puts methodology at center stage. The stories highlight a variety of methods including knowledge elicitation (chap. 2), laboratory studies (chaps. 3 and 8), naturalistic observation (chaps. 7 and 8), usability (chap. 6), and modeling (chap. 5). Perhaps most instructive to the prospective cognitive engineer and interspersed throughout the book, but especially in chapter 9, are the real-life struggles of cognitive engineers to push the science and technology out the door of the lab and into the world where it can make a difference.

We would like to thank the many people who contributed to this book. We especially want to thank the cognitive engineers who are the heroes and heroines of these stories and who were kind enough to share with us the details of their stories and their perspectives on the field: Mike Atwood, Karlene Ball, Frank Drews, Wayne Gray, Bob Helmreich, Bill Howell, Sue Hutchins, Bonnie John, Joan Johnston, Richard Kelly, Dave Klinger, Dick Pew, Eduardo Salas, C.A.P. Smith, Jim Staszewski, Dave Strayer, and Dwayne Westenskow. They are special scientists and special people. Special thanks to William Howell for his commentary. Thanks also to Anne Duffy of Erlbaum who was exceptionally patient and understanding with us. Thanks to Robert Hoffman and the anonymous reviewers of the manuscript. Last, but certainly not least, we would like to thank our partners, Steve Shope and Kate Bleckley, for their support and tolerance in listening to "just one more paragraph," and our children, Jackie, Michaela, Alexis, and Andrew. May they have their share of success stories.

Authors

Nancy Cooke is a professor of applied psychology at Arizona State University at the Polytechnic Campus and is science director of the Cognitive Engineering Research Institute in Mesa, AZ. She is also editor-in-chief of *Human Factors*. Dr. Cooke received a B.A. in psychology from George Mason University in 1981 and received her M.A. and Ph.D. in cognitive psychology from New Mexico State University in 1983 and 1987, respectively. Her research interests include the study of individual and team cognition and its application to the development of cognitive and knowledge engineering methodologies, human–computer interfaces, homeland security systems, remotely operated vehicles, and emergency response systems. In particular, Dr. Cooke specializes in the development, application, and evaluation of methodologies to elicit and assess individual and team cognition. Her most recent work includes empirical and modeling efforts to understand the acquisition and retention of team skill and the measurement of team coordination and team situation awareness especially through the analysis of communication. This work is funded primarily by the Air Force Office of Scientific Research, the Air Force Research Laboratory, and the Office of Naval Research.

Frank Durso received his Ph.D. from State University of New York at Stony Brook and his B.S. from Carnegie-Mellon University. He is professor of psychology at Texas Tech University on the faculties of the human factors program and the applied cognition program. He currently is president-elect of APA's Applied Experimental division, chair of the Aerospace Technical Group of Human Factors, and executive board member of the Society for Applied Research on Memory and Cognition. He was president of the Southwestern Psychological Association and founder of the Oklahoma Psychological Society. A fellow of American Psychological Association and Association for Psychological Science, he serves on the editorial board of *Journal of Experimental Psychology: Applied, Human Factors, Air Traffic Control Quarterly, and Cognitive Technology* and is senior editor of *The Handbook of Applied Cognition*. He is recipient of the Regents' Award for Research and the Kenneth E. Crook award for instruction from the University of Oklahoma, where he served as professor and founding director of the Human Technology Interaction Center. He has been funded by the National Science Foundation and the Federal Aviation Administration. His research interests have focused on cognitive factors in dynamic situations, in particular air traffic control.

Chapter one

Background and perspective

There has been much written about disasters such as the Three Mile Island nuclear accident and the Space Shuttle Challenger explosion and perhaps less about the driver who mistook the accelerator for the brake and the nurse who prepared the wrong drug. However, in all these cases, commentators look to see some human-linked contribution to the tragedy. Recognizing that there is a problem in a human–technical system is the first step, but what comes next? First, are we really clear in understanding the problem? In a blame game, is it the human's fault or is it a consequence of the human–machine context of which the human is just one part? Second, what can we do about the problem? How can we ensure that the disaster does not happen again or at least minimize the chances of its recurrence? Perhaps because the solutions are not as dramatic as the problems, or perhaps because they simply go unnoticed, we rarely hear about this side of the story.

If we gave it some thought, we would notice that the version of the device, software, or vehicle we use today is easier, more flexible, or safer than the one we used 5 years ago. Sometimes we notice, sometimes we do not. And sometimes we cannot notice because, to us, it has always been that way. Younger readers may be surprised to learn that the brake light in the middle of the rear window was not always there or that there was not always a yellow light in our traffic signals. To sink into the oblivion of common practice is success indeed. Consider your notion of how doctors and nurses interact in modern surgery. Readers of all ages have lived in a time when doctors called for an instrument and a nurse slaps the instrument into the doctor's hand. Yet before Frederick Winslow Taylor and Frank and Lillian Gilbreth, forerunners of modern human engineering, had provided the solution, physicians would hunt for the scalpel amid a jumble of instruments, blood, and clutter. How could they not have recognized this simple solution? Ah, the benefit of hindsight!

Most industrial tasks require human operators to interact with various technologies. Unlike 100 years ago, the tasks we ask operators to perform today are highly cognitive, the technologies sophisticated, and the interactions among humans, teams of humans, and machines are highly intricate. The study of cognitively complex human–technical systems falls in the discipline of cognitive engineering sometimes called *cognitive systems engineering*. In this view, a system includes the human worker, the technology, and the interaction between them. This system is not isolated, but part of a broader context. Perhaps the system can be improved by providing better training to the human or using better methods to select the worker in the first place. Perhaps the system can be improved by upgrading or replacing the technology. Perhaps the system can be improved by enhancing the ease with which the human interfaces with the technology. Finally, perhaps the human–technical system is fine, but the context—the social, political, or cultural setting in which it is embedded—is flawed.

Cognitive engineering is the study of cognitive work in context for the purpose of improving system effectiveness and the safety and productivity of the human constituents of the system. Cognitive engineering is part of the science of the field called *human factors* in the United States or *ergonomics* in Europe. Everyone reading this book has heard of the ergonomic chair where human factors researchers have modified the chair to improve its safety, comfort, and aesthetics. The ergonomic chair with its five legs is certainly a success story, but it is a physical ergonomic success. Physical ergonomics attempts to adjust the system by taking into account the physical capabilities of the human operator—the suggested weight limit to lift, the dimensions of the phone line repair bucket, and the shape of the mouse to prevent carpal tunnel syndrome, to mention others beside the ubiquitous chair. Cognitive ergonomics attempts to take into account the cognitive capabilities of the human operator—the amount of information we can expect the operator to remember, the mental diligence to monitor an automated power plant, and the structure of the checklist to prevent a cockpit mishap.

Taking the word *cognitive* to mean thought or mental activities is sufficient for our purposes. Understanding that the word *system* emphasizes the interconnectedness and complexity of modern human–technical interactions is also sufficient for our stories. Perhaps not surprising, exactly what qualifies as a "cognitive system," or even simply cognitive, is easily debated among scientists. However, these are debates of nuance. Cognitive engineers agree on much.

To understand the value of thinking of an entire system, consider how errors occur in complex systems. The popular press often speaks of "human error," as if the only problem were the human operator. Cognitive engineers, instead, see error as emerging within and across layers of the system, with the human being only one of those layers.

James Reason has advanced a great metaphor to capture how errors emerge. Reason's metaphor is sometimes called the *Swiss cheese model* (see Fig. 1.1). Imagine each of a system's layers—from the equipment, to the human, to the work environment, to the coworkers, to the company's culture—as being a piece of Swiss cheese. Continue this metaphor by imagining that a hole in the cheese is a problem. If I take one piece of cheese, full of holes, and add another piece on top, the number of holes running through the cheese is reduced. If I add another piece and another piece, the number of holes all the way through is reduced further. In this view, notice that no one layer of cheese causes the disaster. The human may be partly responsible, but only because he or she is working with equipment that is partly responsible in an

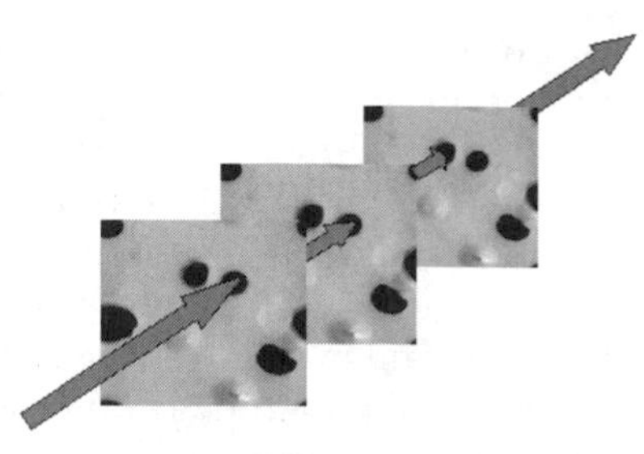

Figure 1.1 According to Reason's model, disaster occurs only when the holes in the Swiss cheese line up. In this situation, the layers of defense have been breached.

environment that is partly responsible. In fact, Charles Perrow has pointed out that even good old-fashioned luck can be an effective piece of cheese. If the pieces of cheese line up and the vehicle swerves into the oncoming lane, whether a disaster occurs will depend on that last piece of cheese—whether another vehicle just happens to be in that lane. This factor is what air traffic controllers mean when they say "it's a big sky" and what doctors mean when they refer to "the resilience of the human spirit."

With the proliferation of modern technology and the complex situations in which these technologies are needed, it is easy to find examples of disasters involving some aspect of human cognition amenable to cognitive engineering solutions. Recognizing that the definition of cognitive engineering is not fixed, but is instead a bit amorphous like the real-world problems it addresses, we believe that stories of cognitive engineering solutions provide a good way to describe the field of cognitive engineering to the reader outside of the area. It is our objective to describe cognitive engineering by way of these stories.

But why do we need cognitive engineers at all to solve such problems? Why not turn to experts who have worked for years in that job? Who better to locate the problem and propose a solution?

First, an unaided subject matter expert may find it difficult to reflect on devices used every day. That is, they will have difficulty distinguishing between routine or required procedures and valuable procedures. They are necessarily focused on their role and their perspective. Second, people, even experts, do not have privileged access to their thoughts and reasons. Research has shown that people will not necessarily be able to tell why they made a decision. The famous education scholar, K. Patricia Cross (1992), asks us to imagine what would happen if we asked "expert" icebox users in the early part of the 1900s to suggest improvements. The answer would have been more ice, more often, not chemical refrigeration. Although it is wise to work closely with specialists in the area, specialists alone are ill-suited to evaluate their own domain. But why are cognitive engineers the conduit to success?

Although anyone can observe, cognitive engineers understand the rules of scientific observation, attempt to be unfettered by bias or at least aware of the biases, and can draw on training and experience to direct observation to the aspects of the situation most relevant. Cognitive engineers use their knowledge of cognition and training in observation to filter the complex barrage of information: Not everything can be observed, so a cognitive engineer observes with a sort of cognitive engineering filter in place—a filter that has been shown to be effective in isolating problems. These observational skills and the development of the right filter take training and practice.

Cognitive engineers are trained in methods and techniques in addition to observation (e.g., interviews, quantitative methods, modeling, and process tracing) that can be brought to bear on the problem as necessary. They have generally spent significant time interacting with and studying people in a variety of situations related to the one at hand. They have an appreciation for a wide range of behaviors, opinions, and perspectives, not merely the one that comes with being an experienced user.

Although there are a number of cognitive engineers successfully helping people, companies, and governments every day, and many instances of positive outcomes, only seven stories appear in this book. We call the examples in this book *cognitive engineering success stories* because they were selected to exemplify the best of what our field currently has to offer. We were strict about our criteria for acceptance as a success story. First, there had to be a relatively dramatic incident, accident, or disaster that created the need for a cognitive engineering solution. This also meant that human cognition had to play some role in the problem.

Next, we required that there was indeed a solution proposed that would necessarily address one or more of the cognitive problems. In our search for success stories, we found that there is no end to proposals for solutions. The next two criteria were by far the most difficult for nominated stories to achieve.

The solution had to be implemented in some way in the field. This means that the multitude of solutions to problems that had been tested in the cognitive engineering lab, but not fielded, were ineligible for inclusion in this book. As you will learn through several of these stories, there is often a gigantic chasm between the laboratory-tested solution and implementation in the field. All kinds of social, cultural, political, and economic considerations need to be addressed, and cognitive engineers are not necessarily equipped to address them. For our purpose, if the solution was never implemented, then it stopped short of being a real solution to the original problem.

Finally, and probably most important, there had to be some evidence that the implemented solution was indeed a solution. We looked for pre–post, before–after evidence that this fix was truly a success. Otherwise ideas that look successful in the lab may for various reasons make no difference in the field or even (when joined with all of the other existing and highly interdependent systems) make matters worse. The solutions that came with solid evidence of success in the field were so rare that we nearly gave in to this last requirement. But in the end, we selected seven stories that were successes by these criteria. In addition to communicating to the general public the field of cognitive engineering, we feel that we have been successful if the nature of our selection process encourages cognitive engineers to push to get their solutions into the field and to collect data that speak to their success. To borrow from therapy development in clinical psychology, we need more evidence-based cognitive engineering.

The seven stories in this book differ not only in the nature of the failure—in the disaster or problem—but also in the approach to the solution and difficulties encountered along the way. It is our intent to represent through these stories a sample of methods, approaches, and theories from the field of cognitive engineering, as well as a sample of the domains to which cognitive engineering has been successfully applied.

Our first story comes from the military. The Staszewski–Davison story about land mines is exemplary of a true success in cognitive engineering. The design of a training program based on a deep understanding of expert performance typifies the cognitive engineering approach. The story also scores high on the elements that we judge to be critical to success: (a) The problem is important (in this case, errors can result in loss of life or limb), (b) there is a training solution proposed, (c) the training has been implemented in the field, and (d) detection rates under the new training regime improved dramatically. Implementation of the new training program was perhaps the most challenging to achieve, but at the same time probably the most rewarding to the research team whose goal was to make a difference in the lives of the soldiers.

The story, "Not Too Old to Drive" (chap. 3), recounts a series of laboratory studies that uncovered a specific perceptual deficit common to older people that affects their driving and can be corrected through training. Some might argue that this is not cognitive engineering, but is about eyesight or pure sensation. However, it is not about eyesight at all, but rather the perception and attention behind it. Further, the fact that it is one person interacting with a motor vehicle makes it appear a simpler system in comparison to nuclear plants and battle ships, but we would disagree. The highway system, with all of its drivers, vehicles, signage, and technology, is a highly complex system and one that most of us come in contact with every day. This story is interesting from a psychological perspective because it demonstrates that a basic cognitive skill can be trained, but its bigger success is in the societal implications it has for simultaneously preserving the mobility of our elderly population while enhancing driving safety.

In our next story, "Get This ... on the Ground " (chap. 4), we leave the roadways to follow University of Texas Professor Bob Helmreich on a sojourn that will take him from studying aquanauts on the ocean floor to crews of modern airliners. Crew resource management (CRM) has evolved over the years and is now required both nationally and internationally. As the story indicates, some of the best evidence for the effectiveness of CRM comes from pilots who have saved lives. In fact, there is promise that such procedures will soon find their way into improving performance of other disciplines, like surgical teams. The Helmreich story highlights the value of rigorous observation coupled with creative analysis. It also shows how the expert in cognitive engineering can develop a program that seems revolutionary to the public, but to the insightful researcher is simply the logical next step. Finally, the evolution of CRM points to the challenges caused by differences among people and cultures.

Our next story, "Number Please" (chap. 5), retells the story of one of the most cited success stories in cognitive engineering. It is the retelling of a story that has risen to legend in the cognitive engineering community, although no lives were lost or ever at risk. This is the story of how Wayne Gray, Bonnie John, and Mike Atwood saved the New England telephone company from spending a lot of money to buy a system that would have ultimately lost more money. This story is also an interesting part of our collection because it relies on the methodologies of modeling human behavior to solve the problem. The story meets all of our criteria, although the fact that the solution was to *not* implement the new system may strike some readers as a rather unusual way of meeting our implementation criterion.

"You Guys Better Take Good Care of Me" (chap. 6) tells the story of Frank Drews and the University of Utah research team's efforts to design a new technology to benefit anesthesiologists. The potential role of cognitive engineering is no clearer than it is in improving patient safety in this country. The story of the Utah team combines passion and insight to produce a monitor that allows anesthesiologists to predict dosages more effectively. The story also illustrates how a cognitive engineer's understanding of one system can be used as an analogy to solve a problem in another domain. With cognitive engineering, not every problem in every domain is a new, unique problem. The team's success led to a new display on Graeger monitors.

In "Too Many Cooks" (chap. 7), we take a look at cognitive engineering in the nuclear industry. Seldom in human factors or cognitive engineering have recommendations of a cognitive engineering expert been accepted and implemented in their entirety. Rather, in most cases, a success story cannot be realized because the promising findings and recommendations never make it to the field. Machines are not redesigned. Systems are not altered. Training programs or materials seem written in stone. Dave Klinger, the cognitive engineering star of this success story, managed not only to quickly identify 50 places for improvement in a nuclear plant, but also managed to get all 50 suggestions implemented and with highly successful results—and U.S. Nuclear Regulatory Commission scores for drill performance were higher after the changes were made.

In our final story, "Decisions at Sea" (chap. 8), we return once again to the military. The Vincennes incident on July 3, 1988, along with the Tactical Decision Making Under Stress (TADMUS) program that followed from 1990 to 1999, have together represented the most frequently cited case in our hunt for success stories. It is the first example to pop to mind for many cognitive engineers. It is indeed an example of a success *par excellence*, meeting all of our criteria for success in a big way. The disaster cost the lives of 290 individuals. The research programs that laid the path toward solutions spanned the two coasts of the United States (San Diego and Orlando) and included dozens of Navy investigators, along with a larger cast of contractors. The solution that was implemented was a set of solutions leading to a cultural change, and the evidence of success is far-reaching. Although much has been written about the disaster, much less has been written about the success of the TADMUS program solution and the factors that led to this success.

However, because of its scope, the TADMUS success story is not like the other stories in this book. Shipboard command-and-control occurs within a large complex sociotechnical system with large complex problems, much larger than human landmine detection or anesthesiologist displays. There was not a single problem, but multiple problems and many potential paths toward solutions. The problems were not solely design or training problems, but clearly required both approaches. Design interventions were needed for individual displays and for the information-processing requirements of the larger system. Training research was needed as well as training systems research. The systems orientation of cognitive engineering is truly put to the test in this system of systems setting.

Suggested readings

Cross, K. P. (1992). *Adults as learners: Increasing participation and facilitating learning.* San Francisco, CA: Jossey-Bass.

Crandall, B., Klein, G., & Hoffman R. R. (2006). *Working minds: A practitioner's handbook of cognitive task analysis.* Cambridge, MA: MIT Press.

Ericsson, K. A., Charness, N., Feltovich, P. J., & Hoffman, R. R. (Eds.). (2006). *Cambridge handbook of expertise and expert performance.* New York: Cambridge University Press.

Perrow, C. (1999). *Normal accidents: Living with high-risk technologies.* Princeton, NJ: Princeton University Press.

Reason, J. (1990). *Human error.* Cambridge, MA: Cambridge University Press.

Chapter two

Harnessing landmine expertise

It was a beautiful Tuscan morning in the fall of 1944 when Jack Wack, a World War II U.S. army combat engineer, grabbed his gear and headed out to the work site in Florence, Italy, with five other combat engineers. The team in the back of the 6X6 truck was headed to Florence to do some work on a bridge that had been blown up. The dozier operator on the team needed to dig some dirt for the project, but was worried because yesterday's dozier and driver had been tragically blown up by landmines. The operator asked Jack and a buddy from Fredericksburg, Texas, to clear another 50 feet.

The two began to clear the area, Jack with the detector and his buddy from Texas 10 feet behind him. They were extra cautious today, not only because of yesterday's tragedy, but because the dozier driver had already set off a mine that had been missed in the previous sweep. If one had been missed, there could be more.

There was a loud "boom," and in an instant Jack's life changed. Landmine survivors talk about their lives becoming instantaneously separated into two distinct lives—the life before the landmine event and the life after. Jack had been tossed into the air by the exploding mine and came back down on his head. His buddy was thrown backward and ended up losing a finger, but was in pretty good shape otherwise.

As Jack lay there, he noticed a large hole in the ground with smoke rising from it and then came a series of thoughts. He was alive and for this he was grateful. He then realized that he had lost a leg. No, make that two legs. Other thoughts rushed through Jack's head. This would be his last mine sweep. Remarkably, he even had thoughts that put a positive spin on this utterly negative situation. Now he could fly home and avoid the sea sickness that he encountered on his trip to Europe!

The dozier operator ran through the minefield to Jack, looked down, and said, "You are OK." Jack was carried by his teammates through the minefield and eventually to an ambulance that carried him to a British field hospital. This was the first stop in a long series of hospital visits over the next 2 years.

At the field hospital, Jack's stumps were cleaned and wrapped in air-tight bandages. He was on his way home.

Although tragic, Jack's landmine encounter and story of survival is not uncommon. The mines remain for years sometimes after a war is long over. They are emplaced because they constitute an extremely effective tactical weapon. In fact, mines are the most feared weapons that a ground soldier faces. Colonel David Hackworth describes in horrifying detail the extent of the fear and the tragedy of the consequences in *Steel My Soldiers' Hearts.* He points out that, "a bullet makes a hole, a chunk of shrapnel may take off an arm—but a mine turns a soldier into a splattered, shrapnel-punctured basket case." Hackworth also tells of battalions in Vietnam who suffered roughly 40% casualties without ever coming face to face with the enemy in combat. Soldiers generally preferred death in combat to being decimated by a landmine.

Although the military casualties are extensive and gruesome, the legacy of landmines knows no truce or peacetime and does not discriminate targets. Landmines do not know that the war is over and continue to claim lives and limbs for years after. The Landmine Survivors Network (www.landmine survivors.org) lists countless survivor stories, including that of Jack Wack, but also includes stories of victims such as Dr. Ken Rutherford, a bilateral lower leg amputee, who was maimed when his vehicle drove over a mine while on a humanitarian mission in Somalia. There is also the story of Jerry White who, at the age of 20, lost his right leg to a landmine in an unmarked minefield while hiking with friends in Israel. Fields that are heavily mined in wartime kill and maim the soldiers for whom they were intended, as well as those who remove mines long after the war and countless innocent civilians.

There are more than 70 mine-affected countries that are the unfortunate hosts to over 60,000,000 live mines. Afghanistan, Angola, Cambodia, Chad, Chechnya, Colombia, India, Iraq, Mozambique, Myanmar, and Sri Lanka have some of the largest numbers of landmine survivors. Landmines kill or injure between 15,000 and 20,000 people annually. One third of landmine incidents result in death. One third to one half of the victims are children.

The presence of landmines has consequences in addition to human injuries and fatalities. They have socioeconomic consequences, in that they greatly inhibit the rebuilding of war-torn communities and economies. Not only are there extensive costs in prosthetic and medical care for the injured, but also agricultural land is rendered useless and countless livestock are destroyed in some of the poorest countries in the world.

Every year, although tens of thousands of mines are removed, it is estimated that many more (e.g., 30 times as many) are emplaced. This imbalance is partially explained by the fact that it costs much more ($300–$1,000) to remove a mine than to emplace it (as little as $3). Other estimates grimly suggest that, without improvement in detection technology or practice, it will take 450 years to clear 45,000 to 50,000 mines.

Landmine detection requires humans using hand-held equipment, such as the AN/PSS-12 metal detector manufactured by Schiebel Corporation, which uses electromagnetic induction (EMI) to detect conductive objects

Figure 2.1 U.S. Army Combat Engineer with PSS-14 near Baghram Airport, Afghanistan, April 2004.

(i.e., metals), or the more recent PSS-14 (see Fig. 2.1), which uses not only induction, but also ground-penetrating radar to detect radar reflections from objects that have an electrical discontinuity from the surrounding media (i.e., soil). Unfortunately, no autonomous technologies are currently available that match the detection capabilities of a human with the hand-held EMI on technology. This is not to say that the human detection rate is exceptional. In fact, the detection rate through the late 1990s was especially poor (e.g., less than 15% probability of detection) for antipersonnel mines with low metallic content such as the M14s. Cognizant of the detection technologies, mine manufacturers purposefully minimize metal components in the mine and use plastics that mimic electrical properties of soil. Detection is made even more difficult in situations in which there is significant metallic clutter or "noise" in the field. This high miss rate is particularly troubling when one examines the tragic life-or-limb consequences of a single error. As researchers Jim Staszewski and Alan Davison put it, "few tasks punish human error as swiftly and savagely as mine detection."

A blueprint for success

Low detection rates, coupled with the high cost of a missed mine, provided a grim backdrop for the research team of Jim Staszewski and Alan Davison. In the spring of 1998, the Army Research Office funded Staszewski, of Carnegie-Mellon University's Psychology Department, to pursue this problem with regard to the contribution of operator knowledge and skill to mine-detection performance using the PSS-12. Staszewski is a cognitive engineer with a background in the workings of study of expertise, especially in the area of memory skills. Staszewski was equipped not only with an understanding of cognitive skill in general, but also with the methodological skill needed to uncover the specific mechanisms underlying cognitive skill in a domain such as landmine detection. Staszewski's proposal and subsequent grant in this area are significant because they indicate that the Army (specifically, Jim Harvey and Dick Weaver) considered avenues for system-wide solutions to this problem (i.e., the human element and the human–machine interface), venturing beyond the frequent, but more compartmentalized, focus on improving machines.

In 1999, a series of meetings was convened by the Army Program Manager of Mines, Countermines, and Demolitions. The purpose of the meetings was to review and evaluate the disappointing performance of the early prototypes of the PSS-14, the next-generation dual-sensor detector (Hand-held Stand-Off Mine Detection System [HSTAMIDS]). The meetings brought together a "Red Team" of experts—individuals from government, military, and private sectors. Staszewski was invited to attend these meetings on the basis of his Army-funded project. It is there that he met Lieutenant Colonel Alan Davison, who had retired from the Army and was operating the Army Research Laboratory field unit at Fort Leonard Wood in Missouri. Prior to the meeting, Davison had attended tests of detector equipment and procedures in Yuma, Arizona, and had briefed the Army Program Manager on various relevant human factors concerns. Davison ended up chairing the human factors subgroup at the meeting. After the initial Red Team meeting, Staszewski and Davison generated some ideas for improved training techniques.

In their discussions, one very interesting fact was pivotal. Specifically, although operators generally had poor detection performance, this was not universally the case. A few operators had significantly higher detection rates of over 90%. These individuals had somehow developed expertise at mine detection using the hand-held EMI technology.

Staszewski, having training in cognitive science and specifically the nature of expertise, viewed this finding as an opportunity. From chick sexing to mental calculations, other research has shown that acquisition of a skill can be accelerated when training is based on lessons learned from the expert. Often experts learn special strategies or develop skills to perceive patterns that enable them to perform at high levels. In many cases, these expert strategies do not match what is taught in the classroom or textbook.

Whether this expertise approach generalized to skills where there was a nontrivial perceptual-motor component was unknown and of great interest. Cognitive engineering provides the technologies and tools to uncover the thought processes, strategies, and skills that the expert applies. Staszewski reasoned that perhaps it would be possible to improve detection probability of the average operator by training operators in the ways of the expert. This approach of bootstrapping the expert's knowledge and skills had primarily been tested and evaluated in the laboratory with relatively simple tasks such as classroom geometry or physics. Although work on applying this approach to richer tasks was ongoing, it was not at all clear whether the approach was ready for complex and high-risk domains like this one.

So Staszewski began his quest to find an expert. Ideally he wanted someone who had lifted several thousands of mines, but who had also managed to avoid injury. He called on various contacts in the government and military. His queries soon led to a retired Army Noncommissioned Officer named Floyd R. Rockwell, a.k.a. "Rocky." Rocky was participating in humanitarian demining efforts and had been involved in demining since 1967. He even helped to test the first PSS-12s. Staszewski contacted Rocky. He had him run through the basic drills and found out that, indeed, Rocky's detection rates were well above average. Other tests were administered to Rocky, as to other experimental participants, to establish baseline data. Rocky was pretty average in other areas, with the exception of his hearing, which was, interestingly, quite poor.

Next, Staszewski applied cognitive engineering methods to get to the bottom of Rocky's success. Staszewski studied Rocky intensively while he was at work detecting landmines. He videotaped Rocky and also recorded the audio output from the detector instrument. He asked Rocky to "think aloud" or to put his thoughts into words as he performed the task. Staszewski then spent hours reviewing the data in the video and the verbal records (called protocols) to analyze Rocky's skills—to better understand what he knew about demining, strategies that he used to detect mines, and specific actions that he took that were relevant to his success. In his analysis, Staszewski found, for instance, that Rocky was using the auditory outputs from the detector to build imaginary spatial patterns on the ground. He would then mentally compare these patterns to ones that he had associated with mines through experience. In addition to this kind of cognitive skill, there were many other interesting cognitive skills that Rocky had developed over time. For instance, Rocky knew exactly how to continually adjust and maintain the fluctuating sensitivity switch on the detector.

Once tapped, this wealth of expert information could then be used to design a training program. This approach of eliciting expertise to train other experts can be thought of as reverse engineering of human expertise. It is an approach akin to that of a paleontologist reconstructing a dinosaur skeleton from a sample of fossils. In other words, "Let's examine some samples of Rocky's behavior, build a model or blueprint of expert behavior from these observations, and use this blueprint to train or build other experts." This approach, a cognitive engineering approach, has been used to build expert systems such as those that make expert-like medical diagnoses and intelligent tutoring systems that instruct based on an expert model. The approach can be contrasted with other approaches to training design, in which instruction is based on the often very good introspection and intuition of the training designers, rather than expert behavior.

Indeed the new expert-based training program was quite different from the training that operators had been receiving. For instance, although traditional training programs taught operators to hold the sensor head 2 or more inches from the ground, the expert blueprint prescribed a procedure whereby the sensor head rested or was slid lightly along the ground surface when the environment permitted—some surface vegetation required shifting to a patting technique. Whereas the classroom training dictated that detection be done a meter at a time, the expert blueprint recommended a foot at most at a time. The new training program, called Cognitive Engineering Based on Expert Skill (CEBES) by Staszewski and Davison, also incorporated plenty of hands-on practice, simplified tests with as much clutter removed as possible, progression to more difficult tasks, and lots of outcome and process feedback.

With the CEBES expert-based training program complete, Staszewski and Davison were ready to take it for a test drive. The first test was done with 22 Fort Leonard Wood soldiers who had just completed Advanced Individual mine-detection training. The tests required some effort to set up. Mine simulants made with hockey pucks, tin cans, and other inert materials were emplaced in a controlled field environment at Fort Leonard Wood for the training and test periods. Glenn Boxley, who was working at Fort Leonard Wood in the Test, Evaluation, and Coordination Office, played a significant part in this and other studies, pulling double shifts to get the job done. Half of the soldiers were given 5 days of the new expert-based training program, and the other half had the traditional training and was used as a control group for comparison.

Did it work?

To rule out the possibility that the new training group was somehow different in initial detection skill levels than the remaining soldiers, a test session was conducted prior to the new training (see Fig. 2.2). Both groups performed at expected levels, with average probability of detecting a mine at about 55% to 58% and the most difficult-to-detect low metal M14 mines being detected only 16% of the time. Hence, all of the soldiers were performing as expected and at fairly low levels of detection.

The real question was how the soldiers with the new training would do once trained. Remarkably, the training on expert knowledge and techniques raised the probability of detection from roughly 57% to 94%. For the most difficult to detect low-metal targets, such as the M14, there was tremendous

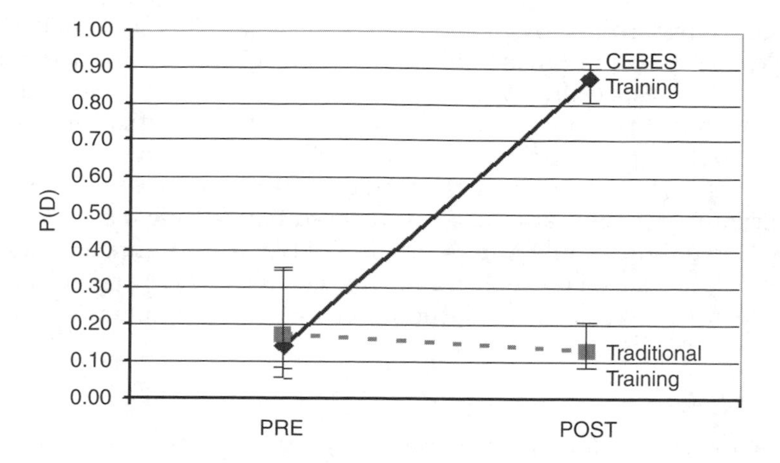

Figure 2.2 Probability of detection of M14s before and after CEBES training.

improvement, with detection probability increasing to over 87%. In other words, the probability of detecting low-metal mines went from 1 in every 6 to 9 in every 10 with the new training program.

The results seemed too good to be true. Perhaps under more realistic conditions the training would not be as effective. What if the soldiers wore the standard body armor purchased by the Army for wear during mine clearing? The research team ran a test with soldiers in full body armor and found similar improvements due to training. Yet what about the detection of actual mines that had been deactivated (i.e., not simulants) under different and more realistic terrain conditions? Again, Staszewski and Davison put the new training program to the test—this time at Aberdeen Proving Ground. In this case, the soldiers who had the new expert-based training had an overall probability of detection of 97%. Half of the time they found all of the mines—100%.

Soldiers were definitely improving their detection skills. Soldiers with the new training detected significantly more mines than those with the traditional training. Thus, there were large differences in what cognitive engineers call *hit rates*. In contrast, soldiers with CEBES training had only slightly more false-positive detections than those without. A false positive is claiming a mine is present when it is not. If the new training raised these types of errors greatly, as well as the hits, it was only making the soldiers more cautious, not more sensitive to detecting the mines. CEBES-trained soldiers falsely identified metallic clutter as a mine slightly less than 1 per every 10 square meters. This rate was not too much higher than the rate for those with traditional training (.5 per every 10 square meters). Clearly, the overall rates for this type of error were well below the military standard of .6 per meter squared (or 6 per every 10 square meters).

Staszewski explains that the new training seemed to change the operators' interpretations of PSS-12 signals in a qualitative way. The audio information was now richer, and now the cue for a mine's location was perceived as a spatial pattern instead of a single audio signal. This seemingly simple change increased the probability of detection. Under the former training regime, mines, especially low-metal mines, were difficult to detect based on the single auditory signal. The new training added a spatial dimension to this signal that greatly increased the probability of detection.

The research team, although thrilled that detection of mines had significantly improved, also realized the need to fine-tune the training program. Notably, they could improve their ability to discriminate mines from metallic clutter, speed up the rate of advance in the process of detection, and examine other conditions of detection such as adverse weather.

Moving to application and large-scale adoption

Despite the overwhelmingly positive results, to have a true impact on countermine operations, the training program needed to be accepted and implemented by those in the field—in this case, the Army. Although the results seemed strong, the research team was faced with some skeptics in the Army who did not believe the data. Therefore, they had little faith in the success of a new training program. It is quite common at this stage for cognitive engineers to face the most resistance to any proposed improvements. Many good ideas do not see past the lab walls. There are many political and economic issues at play here. Often the hesitancy boils down to resistance to change. Change requires more resources (time and money), and it is often presumed that the benefits brought by the change are not worth this cost. Therefore, a strong case has to be made for the change, and this case is further strengthened by salient examples of the adverse consequences of error or system failure, quantitative proof of the idea's merit, advocacy on the part of an influential insider, and pressure from system users.

Staszewski and Davison were persistent in their mission to ultimately help the soldiers. They managed to move their idea outside of the lab. Davison was the influential insider who was not only knowledgeable about the research, but also had the diplomacy and dedication to the soldiers to sell the research results to the Army. The user community was also highly influential in the Army's acceptance of the new training program. Staszewski and Davison were asked to train some Noncommissioned Officers in the field, and soon those trained spread the word. The research team also was funded to train Noncommissioned Officers on how to train troops with the new program. In only 4 to 6 hours of hands-on training, the detection percentage of these trainers increased from 12% to 15% to 80% to 89%, and the word again spread.

In the never-ending pursuit of better-faster-cheaper technologies, the Army decided to scale back the training program to 1 hour per person. The new program had already been demonstrated as superior to traditional training by, in some cases, a factor of six. The cheaper cost criterion had also been met by the relatively low training cost (of under $1,000) to set up a training site for the new training program. Now faster was the only factor that remained. Keep in mind that the initial training sessions in the proof-of-concept experiments took 5 full days—a bit more than the new objective of 1 hour total!

After completing trainer training, Staszewski and Davison held their collective breath to see what would happen. Three weeks later, the commander reported that the reduced training worked well. Word continued to spread about the success of the new training program. All of the elements had come together: adverse consequences of the current training, quantitative proof of success, an influential insider in Davison, and, now, a bevy of users to exert pressure.

Since the original fielding of the training, there have been numerous developments. Most important, the U.S. Army has replaced the old training with adapted expert-based training, as well as the new detection techniques and procedures discovered and tested by Staszewski and Davison. Training was also successfully adapted to a prototype of the hand-held Stand-Off Mine-Detection System (HSTAMIDS). By 2002, soldiers had been given the CEBES training in the use of the PSS-12 and PSS-14 in countermine operations in Iraq and Afghanistan. Training aids were also developed that included the Sweep Monitoring System, the Virtual Mine Lane, the use of mine simulants, a training environment and scoring tools, and a virtual reality–based training system still under development by Carnegie-Mellon University Computer Science Professor H. Herman. Results are also serving as input to new possibilities for semiautonomous detection systems.

By and large, the results generally demonstrate that the expert-based training can be robust under a variety of conditions. Recognizing the critical contribution that operator training makes to HSTAMIDS performance, the Army has adopted the policy of distributing the PSS-14 equipment with the training program as an integrated package.

Lessons learned

In short, the expert-based training is a true success story for cognitive engineering. Significant improvements in mine detection have been made by modeling the training program after the most expert of operators. The story scores high on the three elements that we (the authors of this book) judge to be critical to success: (a) the problem is important (failure to detect can have life-or-limb consequences), (b) there is quantifiable success (human detection rates under the new training regime improved dramatically, and (c) the results of the research have been implemented in the field and have improved the situation. This last criterion was perhaps the most challenging for the researchers to achieve, but at the same time probably the most rewarding to the research team, whose goal was to make a difference in the lives of the soldiers. Reaching beyond these criteria, this approach can be extended to the design of cognitive technologies that augment human cognition and ultimately increase the accuracy and speed of mine detection and removal. Furthermore, this approach is also being applied by other cognitive engineers to a number of real problems, ranging from patient safety in the medical arena to baggage screening for airports.

What became of Jack Wack? He was sent home after the explosion that cost him his two legs and went on to live a full life, with his wife, Judith, eight children, and fifteen grandchildren. He taught engineering at Howard University and worked in the Naval Ordnance Laboratory. Less than 1 month before he died in December 2004, Jack Wack was awarded the Purple Heart.

Suggested readings

Charness, N., Feltovich, P. J., Hoffman, R. R., & Ericsson, A. (Eds.). (2006). *The Cambridge handbook of expertise and expert performance.* Cambridge, England: Cambridge University Press.

Chase, W. G., & Simon, H. A. (1973). The mind's eye in chess. In W. G. Chase (Ed.), *Cognitive skills and their acquisition* (pp. 141–189). Hillsdale, NJ: Erlbaum.

Chi, M. T. H., Feltovich, P. J., & Glaser (1981). Categorization and representation of physics problems by experts and novices. *Cognitive science, 5,* 121–152.

Hackworth, D. H., & England, E. (2002). *Steel my soldiers' hearts.* New York: Simon & Schuster.

Staszewski, J. J., & Davison, A. (2000). Mine detection training based on expert skill. In A. C. Dubey, J. F. Harvey, J. T. Broach, & R. E. Dugan (Eds.), *Detection and remediation technologies for mines and mine-like targets V. Proceedings of SPIE, 4038,* 90–101.

Chapter three

Not too old to drive

The July 2003 day in Santa Monica, CA, was pleasant and uneventful as Joe walked through the Farmers Market on his way back from lunch. He worked nearby and always enjoyed watching the colorful people and activity. It was a Wednesday and market-goers filled the streets, but that was typical because the California market attracted close to 9,000 people every week. There was the after-lunch business crowd purchasing flowers and produce for dinner later that evening. Mothers pushed young children in strollers as they browsed the stands. Thirty-five-year-old Gloria Gonzalez was also enjoying the day. She had decided to purchase some fresh fruit for her two young boys and was working her way through the crowd. A typical day, but no one could have predicted what was to come next.

At 1:47 p.m. and seemingly out of nowhere, a burgundy 1992 Buick LeSabre emerged going approximately 60 mph. The car plowed through the barricades meant to block traffic and kept moving ahead at top speeds, taking down everything in its way. There seemed to be little attempt at stopping as the Le Sabre took down fruit stands and ran over people in its way. By some reports, it seemed to speed up after each impact as if on some intentional rampage through the market. The Buick managed to plow through 1,000 feet of market space within 10 seconds. The car finally came to a stop as a body was tossed up onto the windshield, with another victim trapped underneath it, forcing the final stop. Joe, in amazement, witnessed all of this carnage and pitched in to help 10 people lift the Buick off the woman trapped underneath (see Fig. 3.1).

The driver was 86-year-old George Weller, who looked very confused and disoriented according to eyewitnesses. He claimed that he could not stop the vehicle and later explained that he had accidentally put his foot on the accelerator rather than the brake pedal. It seems clear that Mr. Weller failed to apply the brake and may have actually accelerated instead, but the reason

Figure 3.1 Shortly after the Farmers Market incident.

for this confusion is still unknown. Ten people, including two young children (a 3-year-old girl and a 7-month infant boy), lost their lives that day. Many more were injured. Gloria Gonzalez, who simply wanted to buy some fruit for her children, lay among the dead.

This incident threw fuel onto an already hot controversy and indeed spawned a national debate about elderly drivers and their safety risks. Many angry individuals argued that older people should be given bus passes at the age of 65 and no longer allowed to drive. Still others defended elderly drivers by pointing out the high accident rate (and relative prevalence) of drivers under the age of 25. Some argued for regular testing. Others wondered who to blame. The police claimed it was an accident—the result of human error—although some argued that it was negligence, but on whose part?

Didn't Mr. Weller's relatives know how impaired his driving had become? Weller had renewed his license in November 2000, passing his vision and written tests. His driving record was clean, with only a minor accident. But a deeper look into his driving history revealed that George Weller had recently hit the back of his garage twice while trying to park inside it, and 10 years earlier he had plowed into a wall at a birthday event. There was also much talk about Mr. Weller's seemingly carefree disposition after the incident, perhaps indicating that he was cognitively impaired. These relatively minor events may have gone forever unnoticed if it had not been for the Farmers Market incident. But painting this picture of a history of driving incompetence helped some cast blame on the family or on the state of California, which licensed Mr. Weller. George Weller was released on his own recognizance. His trial was delayed on several occasions, but on October 20, 2006, the 89-year-old Weller was convicted of 10 counts of vehicular manslaughter with gross negligence. He faced up to 18 years in prison, but was one month later determined too ill for prison.

Mr. Weller's case perpetuates the elderly driver stereotype and highlights the issues at hand. We can all list specific incidents of elderly driver accidents or near misses. Indeed, according to National Highway Traffic Safety Administration (NHTSA) statistics, in 1997, older people made up 9% of the population, but were responsible for 14% of the traffic fatalities. Per miles driven, drivers who are 75 years and older have higher fatal crash rates than any other age group, with the exception of teenagers, and the fatality rate for drivers 85 year and older is nine times greater than younger drivers (ages 25–69). Of course, at least some of the discrepancy may be accounted for by the increasing frailty with age. Nonetheless, as the population ages, the percentage of motor vehicle fatalities that older people cause also predictably increases. It is estimated that, by 2024, about 25% of the drivers will be over the age of 65, and we can project that they will be responsible for more than 25% of the fatal crashes.

There is no doubt that elderly drivers are responsible for more than their fair share of accidents. But what can we do about this? Should all elderly drivers be banned from driving? Surely they are not all incapable of driving. Also, when we conceptualize this as an age problem, the prognosis is bleak because there is no correction or training intervention for age. So how do we find the incompetent drivers of the world before it is too late, and what do we do once we have identified them? Can we move beyond simple written road tests and vision screening and get to some of the deeper, perhaps cognitive deficits that may provide more sensitive indicators of driver competence?

By the time of the 2003 incident in Santa Monica, a group of scientists and medical doctors were well on their way to a solution. The first discovery in 1988, which was linked to the ultimate solution, was the realization that the elderly driving problem is not simply due to advanced age or poor vision.

It's not your eyes

In 1988, Dr. Karlene Ball was working with Dr. Daniel Roenker and Dr. Cynthia Owsley in the Western Kentucky University's Vision Lab and the University of Alabama at Birmingham, where they were testing the vision of elderly people. Ball noticed that many of them kept complaining that they were having trouble with everyday activities because of what they thought was their failing vision. They were not only having trouble driving, but also walking and searching for things visually. For instance, they would report that when walking they were surprised by people as if they popped in out of nowhere. However, when there were tested, they did not seem to have any opthalmological or physiological problems.

Ball was intrigued and began to wonder what was at the bottom of these complaints. She gave elderly volunteers many tests; for 5 to 10 hours per person, she measured all kinds of capabilities and skills, looking for a deficit linked to the complaints. Of course finding a relationship or, in statistical terms, a correlation between a measure and a complaint was only the first step. There are many reasons that two things can be related, only one of which is that one thing causes another. Instead, it could be that a correlation between a measure and a complaint was found not because the capability measured (or lack thereof) causes the complaint, but because they are both related to something else (perhaps an age-linked disease) that causes both the deficit and the complaint. A well-known illustration of the distinction between correlation and causation is evident in the positive correlation between tattoos and motorcycle accidents. That is, the more tattoos a person has, the higher his or her incidence of motorcycle accidents. Does this mean that tattoos cause the accidents? Hardly! Instead, tattoos and motorcycle accidents are both linked to a third factor that causes each of them. What is that third factor? It is perhaps mere membership in a particular subculture a sign of a risk-taking personality type, or perhaps both of these hand in hand.

Well aware of the limited interpretations of the correlational evidence, Ball first looked for relationships among capabilities measured and performance and, when promising correlations were found, designed experiments to attempt to move beyond correlation and find evidence for a cause–effect relation. In experiments, one factor is intentionally varied to look for its causal effects on another factor. For experiments to demonstrate this cause–effect relationship, all other factors have to be held constant. Ball, for example, would attempt to correct a weakened capability through some training program and then look for signs of improved performance. The experimental evidence would provide support for the causal link between the factor measured and the problems that the elderly people were having.

All along there were a number of other research challenges for Dr. Ball and her colleagues. Some challenges were associated with using special populations, such as the elderly. Not all individuals are like college students, the *Drosophila* for human performance research. For instance, elderly individuals often require training on aspects of the computer interface that those familiar with computing technology take for granted, such as the use of a mouse. These skills, so basic to the experimenter and most undergrads, can be the cause of some spurious test results if lab personnel assume too much. For the same reason, the researcher needs to be extra cautious when it comes to applying norms or standards from other populations. Ball knew these challenges needed to be overcome to uncover any important deficits.

Out of all the testing—out of all the measurements—came an exciting discovery. It had something to do with how the elderly searched their environment. There was one parameter in a visual search test that seemed to relate to the difficulties the older people were having. This parameter had to do with the size of a person's attentional window, also referred to as *Useful Field of View.*

Useful Field of View is related to the idea of attentional capacity and attentional tunneling or narrowing. Basically, it is the concept that there is only a certain area of the visual field that is available to attention or for rapid use of briefly presented visual information. This is a different concept from peripheral vision, in that a person may have the capability of seeing information in the periphery, but because of Useful Field of View deficits, this information or even information that is visually central is not fully processed to the point that it "registers" and action can be taken.

The UFOV® test, developed by Karlene Ball and colleagues at Visual Awareness, measures the speed at which one can rapidly process multiple stimuli across the visual field. The test involves several parts. The first part displays one of two visual stimuli in a center display box for varying amounts of time. After a masking screen is presented, the object in the box is identified. A second part displays a center visual target as well as a target up to 30 degrees in the periphery. A judgment has to be made about the location in which the peripheral target appears. This decision about where the peripheral object was presented has to be done at the same time as the identification of the centrally presented target, thus adding a divided attention demand to the task. In the third part of the test, the peripheral target is embedded among distracting stimuli. In all three tasks, the screen display is presented for varying lengths of time. Thus, the test captures different aspects of the qualities of the attentional window, including the capability to divide attention between multiple inputs (i.e., the discrimination and radial localization judgments), the ability to selectively process other stimuli in the face of distraction (i.e., the radial localization task embedded in distractors), and speed of processing (discrimination at various display durations). Based on observations of the UFOV® test results for those with the complaints, it was decided that declines of 40% or more in Useful Field of View as measured by the UFOV® test defined significant cognitive decline. After extensive research validating the UFOV® test, it was eventually implemented as a software product by a new company called Visual Awareness, Inc., established in 1988 for this purpose.

But what about driving?

The elderly participants' restricted UFOV® scores were only the beginning of this story. Questions still needed to be addressed: Why were they having this specific problem? What was going wrong? What was causing the restricted attentional window? If the specific cause of this deficit could be determined, then training might be able to target the root of the problem. Maybe the older people were having trouble dividing attention between multiple streams of incoming information. Maybe they were having trouble tuning out irrelevant or distracting information. Still another hypothesis was that they just have generally slower processing. Recall that different aspects of the UFOV® test tapped each of these possibilities.

After additional experimentation, Dr. Ball and her colleagues found that it wasn't any one of those explanations, but rather a combination of all three. Indeed, those individuals who suffered from all three problems appeared to be the greatest risk for a deficit in their Useful Field of View.

Dr. Ball presented this work on the UFOV® test and attentional deficits in elderly people at a National Institute on Aging meeting in 1989. After her talk, she was asked by several individuals at the meeting how this Useful Field of View deficit, so prevalent in older people, related to their driving ability. In her lab, she had looked at a variety of attentional search tasks, but never driving. This was an interesting question and one that she decided to tackle right away.

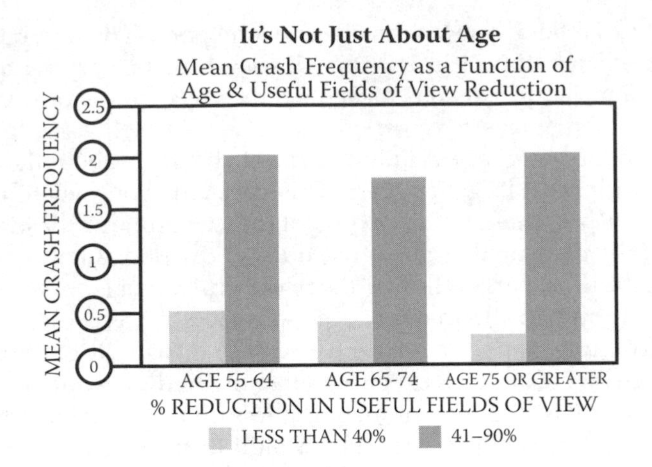

Figure 3.2 Useful Field of View as measured by the UFOV® test is associated with crash frequency independent of age.

In one of the earliest studies, Ball and her colleagues found a relationship between UFOV® scores and driving performance, as reflected in crash statistics for a sample of elderly drivers. Other studies followed, and links were found between the UFOV® test scores of elderly drivers and driving performance as measured not only by crash statistics, but also on-road and simulator driving behavior.

However, you might be thinking that this is a case of correlational evidence, and that correlation between poor UFOV® test results and driving impairment may simply be due to the fact that both are related to aging. However, Dr. Ball's group observed that the relationship was independent of age. The UFOV® test seems to be a bigger predictor of mean crash frequency than even age. In fact, it turns out that poor performance on the UFOV® assessment is associated with poor driving performance across all age groups. However, there is a much greater prevalence of these types of deficits as age increases, with less than 10% experiencing a decline under the age of 65 and more than 40% over the age of 85. Forcing the elderly to take buses would be like forcing people with low verbal scores on their SAT to take remedial math. You would put many people in need of remediation into the math class because there is a correlation between the two tests, but it would be much better to make that decision based on their SAT math score. Targeting people with the triple deficits on the UFOV® test would be much better than targeting all elderly drivers (see Fig. 3.2).

Taking it to the street

The UFOV® test provides a means of screening for problem drivers that is less controversial than screening by age and more sensitive than a simple eye test. As a result, the UFOV® test is gradually making its way into State Departments of Motor Vehicles and self-screening tools such as AAA's "Roadwise Review." The UFOV® assessment is also frequently used by occupational therapists and practitioners in rehabilitation hospitals to evaluate and make recommendations regarding whether someone is fit to drive.

For instance, the Maryland Motor Vehicle Administration (MVA) is currently using the UFOV® test to help physicians on the Medical Advisory Board make judgments regarding fitness to drive. Florida has a "DrivingHealth® Inventory," which is a battery of measures resulting from a research study in Maryland that includes a part of the UFOV® test, and California's Department of Motor Vehicles (DMV) is similarly considering screening with a portion of the UFOV® test as part of their three-tier system.

But the road from the lab to the DMV has been arduous. As Ball recalls, "Everyone thinks that they are experts when it comes to driving. They also 'know' what abilities are related to being a safe or an unsafe driver. After all, they say, isn't it just common sense? Driving is 95% visual." But it really isn't this simple. To be sure, sight distance is relevant to reading road signs, and you cannot drive if you are blind. But when you look at the data, there is a very weak relation between driver safety and visual acuity, a finding that is leading some states to reconsider the visual acuity laws. However, it turns out that cognitive-perceptual problems, like the Useful Field of View deficit, are the more likely culprits when it comes to driver safety.

So why would the UFOV® test be a hard sell to the DMV in the face of the compelling data on Useful Field of View deficits and driving performance? The DMV has a slew of logistical concerns. In some sense, they cannot win. They care about public perception and are attacked if they hint at screening on the basis of age. In addition, they are hesitant to incorporate more testing because then people will complain about the long lines. A further complication is that the states are highly variable in their policies and procedures because every state has its own laws. There are some states (e.g., California, Florida, and Texas) that relicense remotely and will not test for anything, and other states are considering privatized testing, presenting a certificate to those who pass (e.g., California, Colorado, and Nevada). Even when it has been decided that a change should be implemented, it can take a very long time to change public policy.

Although there is constant resistance to using new tests, or any tests, as ways to determine driving competence for licensure, there is a growing trend toward using them to screen drivers for follow-up testing or, in some cases, intervention to mitigate the deficits.

Intervention

The real success story in this case is yet to come. That is, it is truly useful to be able to identify unsafe drivers; however, what to do about the drivers identified is the topic of much controversy. Should the DMV strip the driver of his or her license? Should certain restrictions be placed on his or her license? Alternatively, the passing of the test could be turned into a plus. For example, State Farm Insurance Company is now evaluating a voluntary program in which their insured drivers over the age of 75 can qualify for a discount on their insurance by scoring well on the UFOV® test in a research study in Alabama. This is important for older drivers, who often have specialized insurance with more responsibilities for premiums.

But what becomes of those who do not qualify? Visual Awareness, Inc. developed a training program for those individuals, and it has been evaluated extensively. This is a very interesting development given that these basic perceptual skills used in UFOV® tasks are often considered by some to be untrainable—integral to the basic make-up of each individual. This intervention, for instance, starts with slower, less demanding tasks, gradually increasing demands, and shortening the timeline with repetitions. Trial runs in the lab indicated that this kind of training was effective. Indeed, it served to improve the trainee's UFOV® test performance, as well as other everyday skills. Although a refresher training course seems to be helpful after 1 year, after 2 years, with no refresher, the trainees were still performing the UFOV® test at higher than original levels. The research team also has data which indicate that, compared to simulator training, the UFOV® speed-of-processing training targeting Useful Field of View deficits resulted in improvements on the UFOV® test and transferred to driving performance. This is better evidence for a causal connection between the UFOV® speed-of-processing training and UFOV® test, as well as the UFOV® speed-of-processing training and improved driving performance. Based on these data, we can assume that the UFOV® speed-of-processing training is not just helping drivers pass the test and potentially qualify for an insurance discount, but it may actually be helping the drivers drive more safely. Accident rate data for trainees will provide additional information on the success of the training piece of this puzzle.

A societal success

Beyond the nice story for cognitive engineering, this research has made a significant contribution to a very important part of our society—mobility. Mobility is of special importance in our country, which is largely without sufficient public transportation. There are many drawbacks that come with taking away a person's mobility, including depression, lowered mental stimulation, problems getting around, and, eventually, the need for assisted living. These situations not only harm the individual and his or her family, but also create expenses for society. Clearly, denying mobility to a class of individuals on the basis of age is not satisfactory.

There are also safety issues. What about the death of innocent people who happen to be in the impaired driver's path. Dr. Ball will tell you that if you talk with older adults, they generally acknowledge that there is a problem with "some" elderly drivers even if they do not acknowledge that they have a problem. Most of them are conscientious and safe drivers. For example, older adults often try to modify their environment to remove any increased risk by taking themselves off the road at night or during certain peak-traffic times of day or, notoriously, driving slower.

Dr. Ball and colleagues have offered a solution that stands to increase driver safety while not denying elderly people mobility. They have identified a specific deficit that can be used to identify those elderly drivers at risk, and, even better, they have developed a training program that has been successful at correcting the deficit and improving driving behavior. This solution provides screening that is much better than age or eyesight and, through training, provides hope for those individuals with a restricted Useful Field of View. Furthermore, self-administered versions of the UFOV® test, incorporated in AAA's Roadwise Review, provide elderly drivers with a means of evaluating themselves in the privacy of their own homes so they or their family members can take the initiative to correct their own deficits through training or the modification of their driving behavior. As the population ages and there are more elderly drivers on the road, it will become increasingly important to identify drivers who are at risk and provide them with training or alternatives to preserve safe mobility.

What have we learned since the Farmers Market Massacre of 2003? George Weller could have had a Useful Field of View deficit, he could have had difficulty with sensation in his foot, or it could have been an outcome of senility or many other possibilities. We will probably never know exactly what was wrong with George Weller or many other elderly drivers who have caused accidents and, in many cases, taken the lives of innocent people. However, there is now hope of improved identification of incompetent drivers that goes beyond a vision test and that is not based on age. There is also hope of interventions to improve the driving performance of these identified individuals so they can preserve their mobility without endangering lives.

Lessons learned

There is a chain of successes in this story to consider. First, it was discovered that many elderly people seem to have a perceptual (not visual) deficit involving the speed at which information can be processed in the visual field. The UFOV® test evolved as a means of measuring the extent of this problem. It was then found that performance on the UFOV® test is tied to driving performance independent of age, thereby suggesting one of the factors that may be responsible for driving problems that some, but not all, elderly individuals have. It was then found that training can improve performance on the UFOV® test and transfers to improved driving behavior. This is a terrific example of applied research that starts in the lab and ends up in the DMV. It is also a good example of the value of lab research and the need for correlational, as well as experimental, studies. It was only through well-designed experiments that causal connections could be verified.

Suggested readings

Ball, K., Berch, D. B., Helmers, K. F., Jobe, J. B., Leveck, M. D., Marsike, M., Morris, J. N., Rebock, G. W., Smith, D. M., Tennstedt, S. L., Unverzagt, F. W., & Willis, S. L. (2002). Effects of cognitive training interventions with older adults: A randomized control trial. *JAMA, 288,* 2271–2281.

Ball, K. K., Roenker, D. L., Wadley, V. G., Edwards, J. D., Roth, D. L., McGwin, Jr., G., Raleigh, R., Joyce, J. J., Cissel, G. M., & Dube, T. (2005). Can high-risk older drivers be identified through performance-based measures in a department of motor vehicles setting? *Journal of the American Geriatric Society, 54,* 77–84.

Clay, O. J., Wadley, V. G., Edwards, J. D., Roth, D. L., Roenker, D. L., & Ball, K. K. (2005). Cumulative meta-analysis of the relationship between useful field of view and driving performance in older adults: Current and future implications. *Optometry and Vision Science, 82,* 724–731.

Owsley, C., Ball, K., McGwin, Jr., G., Sloane, M. E., Roenker, D. L., White, M. F., & Overley, T. (1998). Visual processing impairment and risk of motor vehicle crash among older adults. *JAMA, 279,* 1083–1088.

Roenker, D. L., Cissell, G. M., Ball, K. K., Wadley, V. G., & Edwards, J. D. (2003). *Human Factors, 45,* 218–233.

Chapter four

"Get this . . . on the ground"

It was December 28, 1978, and for the passengers on United flight 173, both Christmas and JFK airport were fading memories, their thoughts directed to Portland and the upcoming weekend. A little after 5 p.m., the flight called Portland approach for the first time, "We have the field in sight." The flight had departed from its stopover at Denver 2 hours and 18 minutes earlier with 189 souls on board, including a crew of eight. The DC-8 required about 32,000 pounds of fuel to fly from Denver to Portland, but the plane was filled like dad after the holidays with half again the amount of fuel needed, including the 45 extra minutes required by the Federal Aviation Administration (FAA) and 20 minutes of contingency fuel added by United Airlines.

As the landing gear lowered, the captain noted an unusual "thump, thump in sound and feel." The first officer noted the plane yawed to the right. Although the nose gear light was green, no such assurances glowed from the other landing gear indicator lights.

As United 173 moved toward Portland, Dr. Bob Helmreich, a young professor sat at his desk at the University of Texas, penning a paper for an upcoming conference sponsored by National Aeronatics and Space Administration (NASA) in cooperation with the airline industry. The paper was about the psychology of small groups. Helmreich had studied small groups in high-stress environments since his second year in graduate school at Yale, when he worked with the Navy's aquanauts in Project Sealab in the mid-1960s. Sealab was an effort to study how people worked in pressurized confined spaces on the ocean floor. A few years later, NASA would use these kinds of data to inform its decisions about the Apollo missions. Despite his youth, Helmreich had been known for years for his systematic observational approach to quantifying the behavior of operators under stress when United 173 called Portland approach.

At 5:12 p.m., Portland approach instructed United 173 to contact Portland tower for final landing instructions. However, the suspected problem with the landing gear led the captain to stay with approach control. At this point, the flight had a little over 13,000 pounds of fuel, enough to fly for, at most, 1 hour. Approach control sent the aircraft southeast of the airport so it could stay in a holding pattern within 20 nautical miles (nm) of the airport while the problem was investigated. The crew discussed and performed the necessary checks; the visual indicators on the wings suggested that the gear was down and locked. Twenty-eight minutes from the time the captain notified Portland approach of the possible landing gear problem, he contacted United Airlines maintenance control center, explaining the suspected problem and the steps they had taken. He reported that he now had 7,000 pounds of fuel and intended to hold for another 15 or 20 minutes.

United San Francisco: okay, United 173 ... You estimate that you'll make a landing about 5 minutes past the hour. Is that okay?"

Captain: Ya, that's a good ball park. I'm not goanna hurry the girls. We got about 165 people on board ...

At this point, United 173 had 30 minutes before it would run out of fuel.

Back in Austin, unaware of the peril of flight 173, Helmreich spent those 30 minutes writing about the effect of stress on small groups. Helmreich was no stranger to high-stress living. He not only experienced it personally as part of the U.S. blockade during the Cuban missile crisis, but also when he studied those aquanauts for his dissertation. He worked through the argument that, because attention narrows under stress (like focusing on the landing gear), additional tasks (like overseeing the preparation of the passenger cabin, monitoring fuel) make the situation especially dangerous if the tasks are taken on by the captain. Crew members become more dependent on the captain while the captain becomes less able to monitor the crew.

As if to illustrate Helmreich's point, the captain summoned the senior flight attendant to the cockpit and told her to prepare the passengers and the cabin for a possible abnormal landing.

5:46:52 First Officer [to Flight Engineer]: How much fuel we got ...?
Flight engineer: 5,000
5:48:54: First officer [to Captain]: ...what's the fuel show now ...?
Captain: 5
First officer: 5
Captain: That's about right; the feed pumps are starting to blink
Conversation about landing gear. Heading change from Portland approach. Traffic advisory.
5:50:20: Captain [to Flight engineer]: Give us a current card on weight. Figure about another 15 minutes.
First officer: 15 minutes?
Captain: Yeah, give us 3 or 4,000 pounds on top of zero fuel weight.
Flight engineer: Not enough. 15 minutes is gonna—really run us low on fuel here.
5:50:47: Flight engineer: Okay. Take 3 thousands pounds, two hundred and four.

{aircraft was 18 nm south of the airport in a turn to the NE}
Captain instructed Flight engineer to tell United in Portland that 173 would land with 4000 lbs of fuel. Captain responded affirmatively to landing at 6:05.
Approach descent completed.
5:56:53: First officer: How much fuel you got now?
Flight engineer: 4000 lbs, 1000 in each tank
5:57:30 to 6:00:50 Captain and First officer discuss upcoming abnormal landing.
Report to cockpit that cabin will be ready in another 2 or 3 minutes
{aircraft was 5 nm SE of the airport vectoring to a SW heading}

Helmreich's address to the NASA conference argued that suboptimal management of human resources in the cockpit can have tragic consequences. The industry had to move beyond thinking of pilot error to thinking of crew errors. That meant moving past thinking of technical errors to thinking of communication errors.

Helmreich's aquanauts from years earlier lived on the ocean floor, dropped in water a degree or two away from being ice, where aptly named scorpion fish surrounded the alien intruders from the surface who had to forgo their normal oxygen, toxic at these pressures, to inhale heliox (a mixture of 90% helium and 10% oxygen). In this stressful other-world, Helmreich found that performance correlated positively with the amount of conversation among the team, even when the conversations were back at base, not diving. Interestingly, conversations back to friends and relatives on the surface correlated negatively with performance. The more in-group communication and the less out-group communication, the better the aquanauts performed.

If Helmreich were right, leadership style, crew dynamics, and personality would all be important to the safety of the flying public. Even the culture within the cockpit, the "captain is the captain" mentality, would matter. Until then, no one had argued that entire crews, not an individual, should be the unit of study—crews under high workload, crews in crisis.

6:02:22: Flight engineer: We got about 3 on the fuel and that's it.
Captain: Okay. On touchdown, if the gear folds or something really jumps the track, get those boost pumps off so that ... you might even get the valves open.
6:02:44: First officer [to Portland approach]: ... It'll be our intention, in about 5 minutes, to land on two eight left. ...
6:03:14: Captain [to Portland approach]: They've about finished in the cabin. I'd guess about another three, four or five minutes.
{Aircraft was 8 nm S of the airport on a SW heading}
6:03:23: Captain [to Portland approach]: (We've got) about 4,000, well make it 3,000, pounds of fuel. You can add to that 172 plus 6 lap infants.
6:03:38 to 6:06:10 the flight deck crew prepares for abnormal landing.

6:06:19 Flight attendant [to Captain]: Well, I think we're ready.
{Aircraft was 17 nm S of the airport on a SW heading}
6:06:40: Captain: Okay. We're going to go in now. We should be landing in about five minutes.
First officer [to Captain]: I think you just lost number 1
[To Flight engineer]: better get some cross feeds open there or something
6:06:46: First officer [to Captain]: We're going to lose an engine...
Captain: Why?
First officer: We're losing an engine.
Captain: Why?
First officer: Fuel

At 6:07:12 the captain made the first request for a clearance since the landing gear problem. United 173 was 19 nm SSW of the airport. The last minutes of communication follow:

Flight engineer: We're going to lose number 3 in a minute, too.
It's showing zero.
Captain: You got 1000 pounds, you got to.
Flight engineer: 5000 in there, but we lost it.
Captain: Alright.
Flight engineer: Are you getting it back?
First officer: No number 4. You got that cross feed open?
Flight engineer: No, I haven't got it open. Which one?
Captain: Open I both—get some fuel in there. Got some fuel pressure?
Flight engineer: Yes, sir.
Captain: Rotation. Now she's coming. Okay, watch one and two. We're showing down to zero or a 1000.
Flight engineer: Yeah...
Captain: On number 1?
Flight engineer: Right.
Flight officer: Still not getting it.
Captain: Well, open all four cross feeds.
Flight engineer: All four?
Captain: Yeah.
Flight officer: Alright, now it's coming. It's going to be—on approach though.
Unknown voice: Yeah.
Captain: You got to keep 'me running...
Flight engineer: Yes, sir.
First officer: Get this . . . on the ground.
Flight engineer: Yeah, it's showing not very much more fuel.
We're down to one on the totalizer. Number two is empty.

.

.

.

Flight engineer: We've lost two engines, guys.
Captain: They're all going. We can't make Troutdale (small airport on final approach to Portland).
First officer: We can't make anything.
Captain [to First officer]: Okay. Declare a mayday.
18:13:50: First officer: Portland tower, United one seventy three heavy, Mayday. We're—the engines are flaming out. We're not going to be able to make the airport.

At 6:15 p.m., 3 days before New Year's Eve, the dying United flight crashed 6 miles East Southeast of the airport into suburban Portland, cutting a swath 1,554-feet long and 285-feet wide. The flight engineer, senior flight attendant, and 8 passengers lost their lives. Another 21 people were seriously injured. The aircraft and two unoccupied homes were destroyed.

Helmreich continued his talk with a prophetic wager, not without a tinge of irony. The prediction would ultimately be confirmed in the analysis of the United crash: "I would bet a tank of gas that a significant number of communication breakdowns can be observed under high workload and emergency situations." In fact, the National Transportation Safety Board (NTSB) conclusion indicated that the United 173 crashed with empty tanks of gas because the crew did not communicate effectively about the lack of fuel (see Fig. 4.1).

NTSB analysis would reveal that the landing gear problem was caused by corrosion in the gear, which in turn caused the right main landing gear to fall free. The rapid fall disabled the microswitch for the indicator in the cockpit. Because the left and right landing gear descended at different times, the drag from the right gear caused the temporary yaw that the first officer noticed. Failure to give the flight attendant a time limit to prepare the cabin,

Figure 4.1 The aftermath of Flight 173. It is generally agreed that poor crew resource management was contributory to the crash. The lack of fire damage is due to the absence of fuel at the time of the crash.

as the airline operations manual states, probably failed to convey the correct sense of urgency.

The NTSB attributed the crash to the captain's failure to respond to the fuel state and to the crew's fuel state advisories. The captain was instead focused on the possible landing gear problem and abnormal landing procedure. Contributory was the crew's failure to understand the consequences of the fuel state or to communicate those consequences to the captain effectively.

The board recommended

> Issue an operations bulletin to all air carrier operations inspectors directing them to urge their assigned operators to ensure that their flightcrews are indoctrinated in principles of flightdeck resource management, with particular emphasis on the merits of participative management for captains and assertiveness training for other cockpit crewmembers (NTSB, 1979).

Helmreich had finished writing his talk for the NASA workshop. In hindsight, researchers would look at the workshop as the first conference on cockpit resource management, what is called today crew resource management (CRM). United 173 would, in hindsight, be viewed as the flight that began CRM. In fact, it began in late December both in the interactions of the crew of United 173 and in the mind of Bob Helmreich.

The chief pilot for Texas International (now Continental), J. V. Sclifo, had heard Helmreich's presentation. Helmreich had not proven his position, but he had made a compelling case that made sense on the face of it. Proof would require observations of flight crews. Helmreich knew it. Sclifo knew it.

Sclifo moved toward the podium to congratulate Helmreich on his presentation and, ultimately, to offer valuable help. He gave Dr. Helmreich and his Texas team jump seat access to Texas International Airline (TI) flights. With this kind of access, the human factors researchers from Texas could watch intact flight crews interact in real-world situations. Jump seat access to other airlines followed.

Some of the many jump seat rides are forever etched in memory: Helmreich sat in the jump seat as the aircraft began its takeoff roll. The captain had yet to deploy flaps. Fifty knots, 60 knots, 70. Rotation would occur around 120, and then it would be too late for flaps. Helmreich knew they needed flaps. Should he speak? Would you? After all, he was the guest. 80 knots. Finally, at 90, the copilot said, "Captain, do you want flaps?" resulting in an aborted take-off, but the avoidance of a not uncommon cause of crashes. In another observation flight, the captain of a 727 turned onto the wrong one of two parallel runways, one with another 727 already on it. A final example occurred on a 747 that had pushed back waiting for taxi instructions to begin its flight from Kennedy to Asia. It was a nasty night. The middle of winter. Foggy. Air traffic control (ATC) broke the silence. The ground controller had given a long, complicated set of taxi instructions. The aircraft began to taxi and then stopped in the dark soup of Long Island's weather. After what seemed like minutes, the captain turned to the jumpseat and asked, "Do you know where we are? I can't ask ATC."

From jump seat observations, there followed research time in simulators. More details, more behavioral markers to look for in effective and ineffective crews. It became evident that there was reason to believe that problems in flight crews were real. But what to do about them? The NASA workshop had mobilized the airlines, and a variety of CRM courses had sprung up, with the first at United.

United Airlines launched the first comprehensive CRM program in 1981. Crews participated in a seminar, reflected on their management style, and practiced interpersonal skills in the simulator during simulations of a full flight from preflight briefing to landing and debriefing [a practice called Line-Oriented Flight Training (LOFT)]. The focus was on bringing to awareness issues affecting crew interactions, changing styles, and correcting problems—overassertive captains and underassertive junior officers.

Along with John Lauber, then senior scientist of the human factors program at NASA-Ames and the future first behavioral scientist on the NTSB, Helmreich ran a course on CRM for the check airmen for Texas TI. The offering at TI began with the results of the first full-mission simulation study run at NASA, which identified many issues in communication and decision making, as well as discussion of human factors issues in accidents. The course did not provide guidance for more effective cockpit management. Rather, it was, in today's terminology, a basic awareness program designed to sensitize the check airmen to the importance of the nontechnical aspects of effective cockpit management.

The next few years saw CRM spread to one airline after another. It was not always easy. Many pilots were resistant, and not everyone embraced this "psychobabble," this "charm school." Some aviators even seemed to get worse after CRM training. Some airlines thought it counter to their traditions and philosophy. For example, Delta said it was a captain's airline. CRM would erode the captain's authority, and that was an approach they didn't need or want.

In 1987, Delta Airlines experienced a number of embarrassing incidents, sufficient to attract the attention of the press, including a column in *Time,* entitled "A Case of Delta Blues." Of the six incidents investigated by the NTSB in 1987, five of them pointed to problems in CRM, including limited, misunderstood, or no communication among the crew. In mid-June, a Delta flight mistakenly acted on a takeoff clearance intended for a Southwest flight. Southwest 715 and Delta 314 started their takeoff rolls from opposite ends of the same runway. The Southwest had reached takeoff speed and continued, while the Delta veered off onto a taxiway, narrowly avoiding a head-on collision.

On July 7, a disoriented captain landed a 737 destined for Lexington, Kentucky, at an airport in Frankfort, 17 miles away. Later that same month, a 767 landed on the wrong runway in Boston. The NTSB cited crew coordination and noted that "the Delta captain had a reputation for dominant behavior which tended to suppress others in the cockpit."

On August 2, an L-1011 landed in Atlanta, touching down three times and contacting the runway with the fuselage because of excessive flare (nose up) caused by the captain and the check airman both applying nose-up actions. The final incident occurred in December, when the captain of a 737 became disoriented at LAX after landing and reentered a runway, forcing a United flight to take off over the errant Delta aircraft. The first officer was completing the after-landing checklist at the time.

Delta not only rethought their opinion of CRM, they aggressively embraced it. Helmreich and Lauber, along with J. Richard Hackman, a social psychology professor at Harvard, developed an extensive 3-day CRM course for the airline. Things improved, leading Delta's vice president of flight operations to indicate to Helmreich's team that "things are going too well. We think we changed the culture, but need to validate it." Again, Helmreich and team found themselves in jump seats, flying Delta around the country, looking for behavioral markers of CRM, and relating them to overall crew performance, including errors. The Delta course dealt with specifics of flight operations and focused more on the team and cognition: situation awareness, team building, strategies, decision making, and so on.

So, as it turned out, Delta, "the captain's airline," would become a leader in the evaluation of CRM. In fact, there is now good evidence that programs which include repeated CRM training and practice of interpersonal skills work, although CRM does not necessarily reach everyone.

Although any airline can have a few crewmembers whose behavior does not change or improve with CRM training, overarching cultural influences can stack the deck for or against the effectiveness of CRM for an entire workforce. In other words, exporting CRM to other cultures is not easy. One Asian copilot said, "I'd rather die than question the captain." That airline ultimately flew a Boeing 747 into a mountain with the full cognizance of the junior crew. CRM can have success provided adaptations to the culture are made. One adaptation to the collective culture with great respect for authority was applied successfully: Ask the junior crew to imagine they are the elder son and, as such, have the responsibility to ensure that no dishonor comes to the father.

Courses evolved and extended the concept of crew beyond the cockpit; for example, some courses included joint cockpit–cabin crew training. In 1990, the FAA gave airlines greater flexibility in training in exchange for the requirement that CRM and interpersonal skills training be given and integrated into technical training. In the FAA's advisory circular issued in 1998, it was explicitly stated that "CRM training focuses on situation awareness, communication skills, teamwork, task allocation, and decision making."

As the 20th century moved to a close, CRM, the pedagogy associated with it, and the methodologies used to gather data and evaluate it all continued to evolve. However, problems remained. Not every pilot believed in CRM, the courses differed significantly, and training had to be refreshed every 5 years or so. Some argued that a side effect was a loss of focus on error.

Helmreich sat in his office chatting with his colleague, Ashleigh Merrit, about a rationale for CRM that could be endorsed by all. CRM had been extended, but only in one way. It had been extended to think of the crew as the cockpit officers, the flight attendants, ATC specialists, and so on. However, and this was exciting, it had not been extended to include the environment, the events that precipitated the incidents that CRM was thought to help the crew handle. Bob and Ashleigh bantered back and forth, heading toward another generation of CRM:

> We need CRM because it helps the crew do their job better if they've had that training than if they haven't. But what exactly is that job? The job is NOT interacting well with your teammates. That's important to do the job, but it is not the job. The job is really to manage problems. Problems like threats from weather, equipment, other aircraft, and so on.
>
> And threats of their own making.
>
> Yes. Errors made by the crew become threats that they have to manage just like they have to manage problems not of their doing such as weather and mechanical breakdowns.
>
> So, crews do threat and error management. Once the error has been made or once the threat presents itself, the job is to move from the situation caused by the threat or the error to a safe situation, one without threats, one not caused by an error.
>
> There's no psychobabble in managing threats and errors. Any flight crew would see threat and error management as a respectable job, an honorable job. If having better team communication meant better threat and error management, then crews would see it as valuable.
>
> So, we have to do more than extend the concept of crew to include other aviation personnel; we must extend the concept of the cockpit to include the environmental events that impinge on it. We have to put the cockpit and the crew in context.

Helmreich and his team were getting good data on the crew. What was missing was capturing the context. Significant support would be needed to begin this innovative path. The support was ensured when the head of safety of Continental visited the lab in Austin. As he was pulling into the parking lot at the University of Texas (UT), one of his DC-9s was landing with its gear up.

Helmreich and the UT team revamped the entire observation protocol, developing the Line Operation Safety Audit, and did this in the context of a model of threat-and-error management. Many human factors researchers at the time were focused like a laser beam on the error. To Helmreich, the error was blood under the bridge: "Once you screwed up, how do you manage it?" You can manage it and make it inconsequential if it isn't already, an outcome that research suggests occurs about two thirds of the time. Or the error can lead directly to a high-risk state, like the wrong heading or altitude. One must now manage this high-risk state. Or, rarely, an error chain can begin, cascading into a disaster.

By 1985, six years after Helmreich spoke in San Francisco, CRM had become mandated by the FAA, and by 2005, the Line Operations Safety Audit was mandated in 186 countries by the International Civil Aviation Organization, the United Nations regulatory body for aviation. Although the idea of looking at cognitive and social aspects of teams swept dramatically through the aviation industry, it promises to sweep even more rapidly through health care, process control, and the oil industry. Thus, the interaction of the three men in the cockpit of United 173 in the winter of 1978 has evolved over almost three decades to expand from the cockpit crew to the team of professionals with a stake in the aircraft, and from merely the events within the cockpit to the threats and errors that impact it. Airlines and pilots seem to have moved from a skepticism and disdain for CRM to an acceptance of, and even an appreciation for, the human factor.

Perhaps the best evidence for the success of CRM comes from the pilots. Consider Captain Al Haynes and the often televised crash of United 232 at Sioux City (see Fig. 4.2). The flight lost all hydraulics, making control of the plane by conventional means, the ailerons, impossible. The crew flew and steered the crippled aircraft by using engine thrust. The coordination among the crew saved 184 of the 296 on board. Haynes attributes the success, in large part, to CRM: "I am firmly convinced that the best preparation we had is a program that United Airlines started in 1980 called Command Leadership Resource Management training." He continued, "It is now referred to as Cockpit Resource Management" (airdisaster.com).

CRM entered the aviation community like a karate chop, but what appeared to the rest of us as a rapid, illuminating change in aviation did not arrive to Helmreich as a moment of insight. To Helmreich, the creative idea to look at cognitive and social aspects of teams in the cockpit was simply a matter of following his interests into a new domain. What he

Figure 4.2 Although United 232 crashed, the fact that anyone survived was attributed to effective communication among the crew and with a pilot traveling as a passenger. The photo shows the aftermath of the landing.

learned on the ocean floor he taught to the men and women who fly far above it.

Lessons learned

Expanding from the individual outward to include the cognitive-social milieu of a complex industrial task is a critical lesson learned. Today, we understand that influences on individual performance come from teammates as well as higher order macroergonomic factors, including the company's philosophy, political pressures, and culture. Cognitive engineers have also developed markers that allow trained observers to distinguish a functioning teammate from a dysfunctional one, the skilled from the unskilled, the good from the bad. CRM is a prime example of how a demonstrable success can arise from good ideas taken from one domain and applied to another. Of course, it takes the appropriate human factors training in cognitive engineering for the researcher to recognize the good idea that can be transferred to the new situation, the markers in one industrial task that will be present in the other, and the countermeasure that can be exported from one culture to another. The expansion away from the individual proceeds along other dimensions as well, including to the environment, the task, and, most important, to other safety critical industries like health care.

Suggested readings

Helmreich, R. L., Merritt, A, C., & Wilhelm, J. A. (1999). The evolution of crew management training in commercial aviation. *International Journal of Aviation Psychology, 9*, 19–32.

National Transportation Safety Board. (1979). *Aircraft Accident Report: United Airlines, Inc. Douglas DC-8-54, N8082U, Portland, Oregon, December 28, 1978 (NSB-AAR-79-7)*. Washington, DC: Author.

Reason, J. (1990). *Human error.* New York: Cambridge University Press.

Salas, E., & Fiore, S. (2004). *Team cognition*. Washington, DC: American Psychological Association.

Sexton, J. B., Thomas, E. J., & Helmreich, R. L. (2000). Error, stress, and teamwork in medicine and aviation: Cross sectional surveys. *British Medical Journal, 320*, 745–749.

Chapter five

Number please

Dr. Wayne Gray sat watching Lily Tomlin's character, Ernestine, on a rerun of *Saturday Night Live*. "Here at the Phone Company we handle 84 billion calls a year," Ernestine bragged. "Serving everyone from presidents and kings to scum of the earth." Tomlin had created Ernestine for the 90's television show, *Laugh-in*. Her parody of interactions with the phone company won many fans during those days of the telephone monopoly. In fact, the reputation of rude operators began with the first operators, who were young boys. The boys, when they were not wrestling or throwing spitballs, were more than a match for an outraged caller. "... this phone system consists of a multibillion-dollar matrix of space-age technology that is so sophisticated," Ernestine continued, "even we can't handle it. But that's your problem, isn't it?" Gray smiled. Well, it was his problem now.

It was 1988. A brief long-distance call in the 1980s would cost a little under $3.00 for a coast-to-coast, daytime call if, that is, you dialed the call yourself. The cost of a phone call, even a direct-dial call, depended on the duration that the line was in use and the distance of the call.

The industry has always been sensitive to factors such as time. For example, even our current numbering-plan-areas, commonly called area *codes,* are an artifact of the days when dialing some numbers took longer than others. Original area codes established in the middle of the last century had either a "1" or a "0" for the second digit and were correlated with the population. For example, the area code for Manhattan, the country's most populous area, was 212 because dialing time would be short; Los Angeles, 213; Chicago, 312. Compare those with South Dakota's 605, a digit sequence that would require considerably more time for the dialed number to register. Of course, now with touch-key "dialing," it no longer takes longer to punch a 9 than a 1. Before the introduction of the push-button phone, however, a caller stuck her finger in the dial opening, moved the dial until her finger struck the stop bar, and then, if she were a conscientious dialer, would remove her finger so that the dial could return to its original position at the appropriate speed to allow the number to be registered. It would physically take the longest time to move the dial for 0 to the stop bar and wait for it to return and the shortest time for a 1.

Another factor that influenced pricing was whether the caller needed to talk to a toll and assistance operator (an information-assistance operator was free in 1988). Even 20 years after Ernestine first appeared on TV, interactions with telephone operators had not changed much. If someone wanted to call another number, but have the call billed in an unusual way—such as having the receiving party pay, called "reversing the charges," or "person-to-person," where billing didn't begin until the particular person came to the phone, or calling-cards, or billing to a third number—then they had to dial a 0), which everyone knew stood for operator. For an operator-assisted call, like reversing the charges, pricing was considerably higher. The more time an operator spent on the line with a customer, the more money it cost to place the call.

An operator-assisted long-distance call in 1988 would begin with the customer dialing a 0. She could then continue dialing the number or wait until the operator answered. If a phone number occurred after the 0, the operator would know the destination of the call, but not the billing method. The workstation would beep and the operator would say, "New England Telephone, may I help you?" The caller might then say, "I'd like to make a collect call to Bob Wehadababyitsaboi." The operator would connect the call and, when the phone was answered, would indicate, "Person-to-person call for Bob Wehadababyitsaboi. Will you accept the charges?" Charges, at the higher person-to-person rate, would begin when Bob came to the phone. If the intended recipient of the call were not there, no charges were incurred.

Just last week, Gray's boss of less than a year, Mike Atwood, had given him an assignment that would lead to a project that would prove to be the most successful use of cognitive modeling of real-world performance to date.

What is modeling?

Many people incorrectly think of an architectural or physical model when thinking of a scientific model. A scientific model is not a physical version of the actual system, simply smaller. Baby boomers remember the solar system "model" of the atom that graced every elementary school in post–World War II U.S. classrooms and serves as a symbol of nuclear energy. This (incorrect) depiction of an atom is based on what scientists would consider a model, the Bohr model, but the depiction itself is not a model. The Bohr model, or any model, is much more than a depiction or replica.

In fact, the organization, Modeling for Understanding in Science Education (MUSE), attempts to improve science understanding in the United States in Grades K through 12 by focusing on scientific modeling. A MUSE curriculum differs from traditional curricula in sacrificing broad coverage of facts for in-depth understanding of models. MUSE students are told that a model is "a set of ideas that describes a natural process and can be used to explain a specific set of phenomena" (Cartier, Rudolph, & Stewart, 1999). Although there are more complicated definitions and a variety of ways in which scientists model (and thus a variety of resulting types of models), the definition that serves MUSE students will serve us well.

With a model, the scientist can do more than describe the phenomenon. She can explain why the phenomenon occurs, predict the future state of the system, and, in some cases, even control it to reach a desired outcome. In this way, a model provides an organized, coherent understanding of the phenomenon and what causes it.

Models are especially useful when the phenomena under study are complicated, as is virtually always the case in human–technical systems. Yet this complexity does not necessarily demand complex models. Simple models often produce complex outputs.

For example, the computer-generated orcs in the huge battle scenes of *Lord of the Rings* can be seen engaging in a myriad of behaviors, including some running away from the battle, all based on a few simple operating principles. Consider the flow of automobile traffic. A simple model can account for much. Assume cars move within certain speeds. If there is not much distance between the car and one in front, then the car slows down. If there is too much room, the car speeds up. These simple assumptions can produce dramatically complex patterns. We have all experienced being stuck in a traffic jam. We crawl along at well below the speed limit until suddenly we reach a point that is clearly the end of the traffic jam. Yet as we burst forth from the jam, we see nothing indicating the cause of the jam. The accident that was there is no longer to be seen. All of this complexity can be modeled with a simple set of assumptions.

Working with models

A good model accounts for most of the known characteristics of the phenomenon and many of the inferred characteristics. It provides a better understanding of the system or phenomenon than was previously held. A model is a schematic of the system, theory, or phenomenon. As such, it is simpler than the phenomenon it tries to explain.

Of course, science has other methods to understand phenomena, like experiments. Even with experiments, models provide much in the way of organization and understanding. However, there are times when the only way to understand a system is to model it. There are no other scientific techniques that have been successfully used to understand earthquakes, or for that matter our traffic jam. Models are often cheaper and faster than conducting experiments. In addition, a good model yields good predictions. A model of forest fires would be of little value if it only "predicted" the past forest fires on which it were based.

For a model to work, researchers begin with a set of ideas and interconnections among those ideas. The ideas are based on empirical observations that have already been published or are collected especially for the project at hand. These ideas, together with empirical laws, can be added to other ideas and mechanisms. Yet if the ideas on which the model is based are incorrect, the model will yield faulty descriptions and predictions.

When the model is finally applied to empirical data, the researcher hopes to predict the data perfectly or at least reasonably well (and scientists can quantify *reasonable*). When the model fits, it can be assumed that the ideas on which the model was based have some truth.

A researcher continually interacts with the model, modifying it in principled ways to account for more of the known and inferred characteristics. One model may not account for all of the known characteristics and yet still prove valuable. For example, Newton's model of the solar system was retained for generations, although it was known that it did not account for the details of planetary movement. The modification a researcher makes must be principled—that is, it must be made in ways that are consistent with fundamental laws and understanding of the phenomenon. Making modifications or "tweaks" merely to "fit" a model to a particular instance of the phenomenon does not add to understanding and often produces predictions for other instances that are unacceptable.

There are, in fact, many systems that can be modeled and a number of ways to model them. Chemists make assumptions about the speed of chemical reactions and predict slow processes (like oxidation) and fast reactions (like explosions). Aeronautical engineers make assumptions about fluid dynamics and the structure of aircraft and predict the speed and capacity of modern carriers. Civil engineers make assumptions about structure and stress and predict the capacity of bridges. Astronomers make assumptions about matter and energy and predict the future of the universe. Cognitive engineers make assumptions about thoughts and behaviors and predict the mind.

"Wayne," Atwood said. "They need someone to go to Boston. They probably want help doing their statistics. We do not do statistics."

"Boston harbor has some great sushi," Gray quipped. "You know we have to go."

"Alright, it's yours. Keep me informed . . . that is, if anything happens worth informing me."

It soon became clear to Gray that NYNEX (New York/New England phone company) New England needed help in the design of a field trial for a new operator workstation. Of course, they would need help with analyzing their data, but this could also be an opportunity to create a GOMS model and compare it to real-world data.

A human factors solution

The 1980s witnessed the beginning of serious thinking about applying cognitive psychology and cognitive science to practical problems. Don Norman coined the term *cognitive systems engineering*, Hollnagel and Woods were looking for a cognitively oriented engineering, and Rasmussen's work done years earlier had made its way to widely available outlets. A technique for modeling cognition in complex tasks, called Goals, Operators, Methods, and Selection rules (GOMS), was developed by Stu Card and Tom Moran at Xerox's Palo Alto Research Center (PARC), with Al Newell of Carnegie-Mellon University (CMU). To avoid the obvious confusion between telephone operators and psychological operators, we refer to the psychological operators as OPS.

Gray was well aware of GOMS, having spent a year at CMU before coming to NYNEX. There he had met Bonnie John, a young doctoral student of Alan Newell's, who for her dissertation was trying to extend GOMS to allow modeling of tasks that had several components occurring simultaneously. Gray knew that telephone operators performed many tasks at the same time, and he knew that John's version of GOMS could be the key. He got permission from Atwood to bring John on board the project. John had just defended her dissertation, in which she developed and tested what would become Cognitive, Perceptual, and Motor (CPM)-GOMS. Cognitive, perceptual, and motor components were, for the first time, put together in a modeling system in a principled way. John had used CPM-GOMS to explain most of what psychologists knew about the behavior of skilled typists. Now she was ready to do the same for telephone operators.

The telephone operator's task was a nice fit for the GOMS modeling procedure. There were clear goals, OPS, and methods. Consider the operator's goals. In our call to Bob, the operator had to determine, "Who pays?", "At what rate?", and "Starting when?" Thus, toll operators had three simple goals.

What OPS does she use to reach these goals? She listens, she talks, she reads and writes, and she keys in numbers. Finally, there are the cognitive activities needed to process information and make decisions. Clearly, many of these OPS occurred in parallel. Only John's CPM-GOMS could model that.

A method is the procedure for accomplishing a goal. It is a conditional sequence of goals and OPS, during which it is assumed that the operator checks her working memory and the state of the task environment. Methods are already part of the operator's knowledge. They are not created on the fly. Consider Fig. 5.1, which shows the overall method for completing a call.

Finally, selection rules played virtually no role in this modeling effort. Selection rules normally constitute the control structure used to select among alternative methods for doing the task to reach the goal. The telephone operator's task was so constrained that there are, for all intents and purposes, no alternative methods and, thus, no selection rules.

"They want us to give them a number, Mike," Wayne reported. "How much will the new system save them . . . precisely?"

"Tell me, did they say Number, please?" Atwood smiled.

"The new workstation they're getting puts up a screen of information almost a second faster than the old one, 880 msec if we're being precise," Atwood answered. "And, the new workstation is being structured to require

GOMS Model	Observed Activities
GOAL: HANDLE-CALLS	
. GOAL: HANDLE-CALL	
. . GOAL: INITIATE-CALL	
. . . GOAL: RECEIVE-INFORMATION	
. . . . LISTEN-FOR-BEEP	Workstation: Beep
. . . . READ-SCREEN(2)	Workstation: Displays information
. . . GOAL: REQUEST-INFORMATION	
. . . . GREET-CUSTOMER	TAO: "New England Telephone, may I help you?"
. . GOAL: ENTER-WHO-PAYS	
. . . GOAL: RECEIVE-INFORMATION	
. . . . LISTEN-TO-CUSTOMER	Customer: "Operator, bill this to 412-555-1212-1234."
. . . GOAL: ENTER-INFORMATION	
. . . . ENTER-COMMAND	TAO: Hit F1 key
. . . . ENTER-CALLING-CARD-NUMBER	TAO: Hit 14 numeric keys
. . GOAL: ENTER-BILLING-RATE	
. . . GOAL: RECEIVE-INFORMATION	
. . . . READ-SCREEN(1)	
. . . GOAL: ENTER-INFORMATION	
. . . . ENTER-COMMAND	TAO: Hit F2 key
. . GOAL: COMPLETE-CALL	
. . . GOAL: REQUEST-INFORMATION	
. . . . ENTER-COMMAND	TAO: Hit F3 key
. . . GOAL: RECEIVE-INFORMATION	
. . . . READ-SCREEN(3)	Workstation: Displays credit-card authorization
. . . GOAL: RELEASE-CALL	
. . . . THANK-CUSTOMER	TAO: "Thank you"
. . . . ENTER-COMMAND	TAO: Hit F4 key

Figure 5.1 Activity-level GOMS analysis of the TAO's task. GRAY 1993 HCI V8 237–309. Figure 4, p. 248 Gray et al.

fewer keystrokes," Atwood said, telling Gray things they both knew. "If I remember, a back-of-an-envelope estimate was something like 4.1 seconds per call. At $3 million a second, that should save about 12.3 million dollars a year. So, they don't want back-of-an-envelope. They want it precisely. Big deal. Not much interesting science there."

"We could try a GOMS model. When the field trial is finished, we can compare the model's results to real world data." Although the decision to purchase the new workstation had essentially already been made, field trials were used, not necessarily to collect data although that happened, but to gain in house experience in training, using, and maintaining equipment.

"And," Atwood joined in, recognizing immediately where Gray was going. "In the meanwhile we can give them their number."

"Precisely."

Any data from the field trial wouldn't be available until next summer. It would actually be interesting to model the new workstation and the old one entirely from specifications, Gray reasoned. Then we could compare our models with how well it matched the real findings from the field trial.

How GOMS works

With Dr. John joining the team, they got started. The plan was to build the models while the trial was being conducted.

How does GOMS produce estimates of time saved and, by inference, money saved? For a chosen method, each activity has some time associated with it. For example, it might take one tenth of a second, or 100 msec, to LISTEN-TO-BEEP. It might take 340 msec to READ SCREEN. Modelers can estimate activities in a number of ways. Sometimes information already exists in the literature. Other times psychological equations can predict times. For example, the time it takes to move a finger to a target depends on both the distance to the target and its size and can be computed precisely using an equation called Fitt's Law. Finally, estimates of time can come directly from observing the task at hand. In fact, estimates of all observable OPS on the NYNEX project came from videotapes of calls. Only unobservable OPS, like READ SCREEN, came from prior research.

According to the videotape, LISTEN-TO-BEEP does take 100 msec; GREET-CUSTOMER, 1,570 msec; ENTER-CALLING-CARD-NUMBER, 4,470 msec; and so on. In traditional GOMS, these times would be summed across OPS for the particular method. Gray and John knew that doing this would overestimate most calls because often many activities occurred while the operator was listening to the customer—commands were entered and credit card numbers were input. In fact, ignoring the parallel nature of the task in the diagram produces an error of 4.85 seconds. GOMS would predict that a call depicted in the diagram would cost New England Telephone $14.55 million dollars more a year than was the case. For CPM-GOMS, the parallelism in the telephone operator's task is not a problem.

It was April 1989 when the final CPM-GOMS was completed. Gray and John had taped 12 notebook sheets of paper together, producing a 94-inch scroll that would have been the pride of any medieval decree. Today found Wayne and Bonnie in the executive boardroom, which was just big enough to hold the 4-foot-wide rosewood conference table. The table was the only place that could accommodate the GOMS scroll.

"I don't think they'll appreciate you crawling on the table," Bonnie said to Wayne as he lay on the table poring over the results. "This can't be right," he muttered to anyone who cared to listen. An average phone call will take 0.6 seconds longer with the new system.

Bonnie climbed onto the table. They spent what seemed like hours perched on the table poring over the scrolls. "They're going to lose money with the new system."

"Almost $2 million a year."

They headed to Atwood's office, with the scroll tucked under Gray's arm.

"Mike, the model predicts that the new system will actually take over half a second longer for a typical call than does the current system."

"How can that be? The display is faster. There are fewer keystrokes. Did you doublecheck your calculations?"

Of course they had. The figure shows charts of the current and proposed workstations. Indeed, the proposed workstation required fewer keystrokes overall. But what keystrokes were removed? The proposed workstation took all of the keystrokes from a time when the operator was performing other tasks. These other tasks continued along at the same pace as before. So the operator entered fewer keystrokes, but she still had to wait for the customer to quit talking. Result? No time saved.

Gray and John showed that the keystrokes removed were not on the critical path. The critical path is that set of activities that defines the longest time. So if a customer talks for 6,280 msec while the operator simultaneously enters the calling card number for 4,470 msec, reducing the latter will do nothing to the length of the call because it does nothing to the critical path. This critical path methodology is so critical to CPM-GOMS that the CPM part of the name is often thought of as *critical path method*.

Imagine you are balancing this week's transactions in your checkbook on a flight from New York to Los Angeles. If the passenger next to you loans you a calculator, this technological aid will not reduce the time it takes to get to Los Angeles, although it will help with the financial task. The critical path (containing the flight) was unaffected by expediting a task on a secondary path (the checkbook).

Ironically, although the keystrokes removed by the proposed NYNEX workstation did nothing to the critical path, one of the few keystrokes added by the proposed workstation did fall on the critical path, thus actually increasing the time to make a phone call (see right panel of the Fig. 5.2).

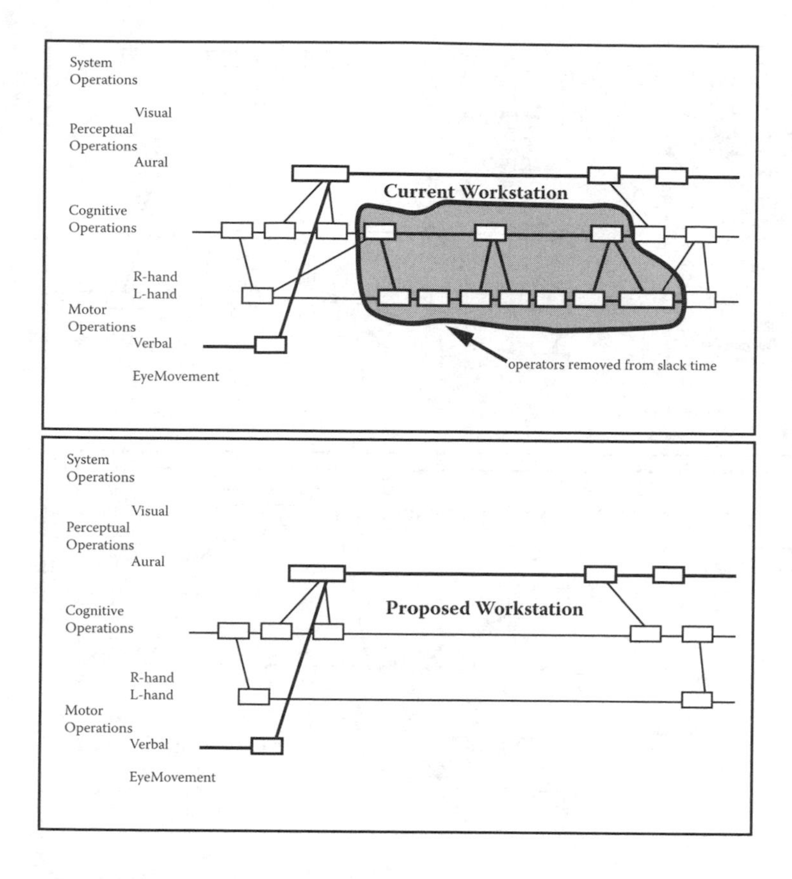

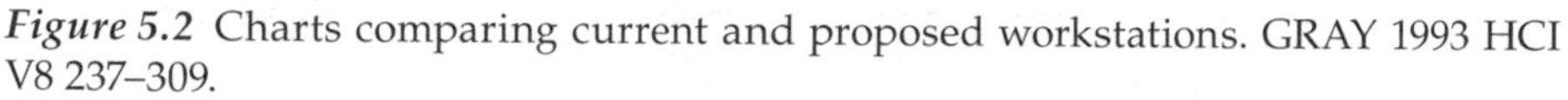
Figure 5.2 Charts comparing current and proposed workstations. GRAY 1993 HCI V8 237–309.

The field trial

The April field trial data were available in late May 1989. The team had completed the model just 2 months earlier. Atwood had alerted people in the company, but his dire answer to their "Number, please?" request was met with, at best, indifference. No one outside of the scientific community really believed the projected negative consequences that GOMS attributed to the proposed workstation. They were content to wait for the field trial.

The field trial delivered 78,240 work times collected over a 4-month period. Twenty-four operators who worked on 12 old workstations were compared with 24 who worked on 12 of the proposed workstations. Each month data rolled in. In the first month, the estimate delivered to management showed a loss. A loss in efficiency of 4%, or about .8 seconds, would cost the company $2.4 million. Surely the operators have not become comfortable with the new system. The next month showed the gap was closing. Relief. By the next month, it was clear the gap would not close. The loss would be real.

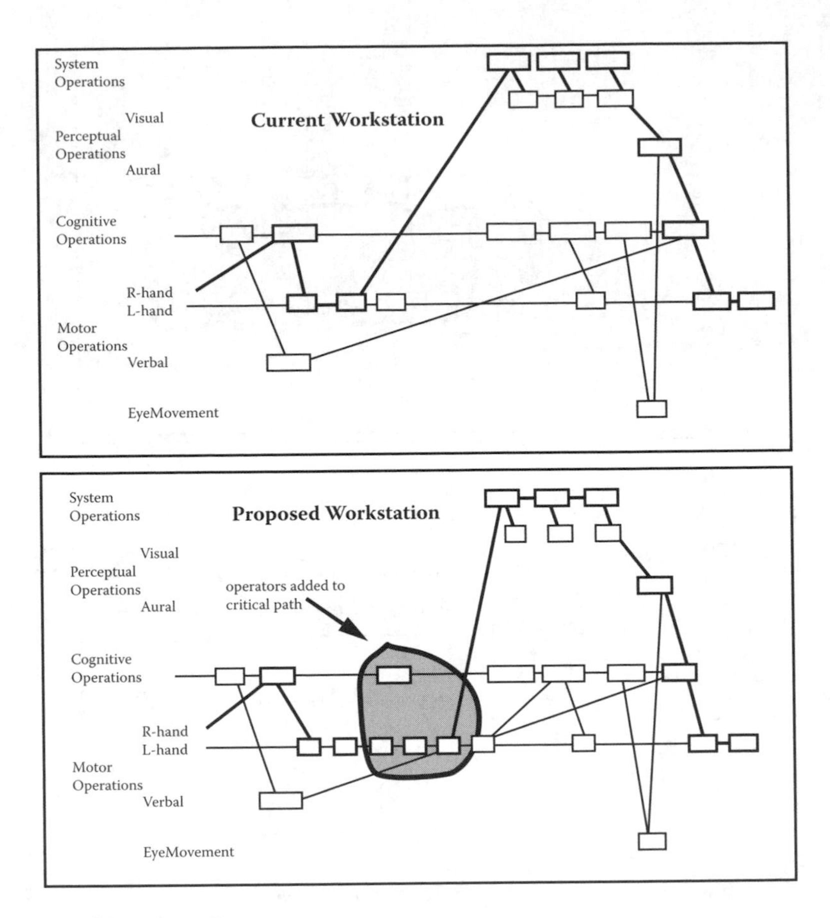

Figure 5.2 **(*Continued*)**

NYNEX's response was understandable. They blamed the trial itself. They blamed the training the operators had received on the new workstation. They blamed the operators. They blamed the equipment—it must be buggy prototypes.

Eventually, management remembered the estimate from CPM-GOMS—.6 seconds slower. No one had believed the model before. Now they turned to it for an explanation.

So, Science & Technology explained to management about critical paths, and about how the proposed workstation took away keystrokes from secondary paths, but actually added them to the critical path. Other large Telcoms adopted the new system. NYNEX never did, thanks to the trial and to CPM-GOMS.

Recently, Bonnie John and a team at National Aeronautics and Space Administration (NASA) (i.e.., Michael Freed, Michael Matessa, Roger Remington, and Alsonso Vera) developed algorithms that automate much of the GOMS modeling process. What once required researchers to crawl across rosewood tables can be done on a more virtual desktop.

Lessons learned

New technologies are usually introduced because they are thought to be better: easier to use, more efficient, and cheaper. But how does one know? In the past, introducing new technologies or introducing new procedures into the human–technical system would require that the new technology be developed, introduced into the work environment, and tested.

The work of the Project Ernestine team pointed out that a project thought to save $12 million would actually cost $2 million. They did this by modeling a system based on specifications, not by waiting for the empirical results. When the empirical data did appear, it confirmed the model (within .2 seconds per call).

Modern modeling, facilitated by the computational power available today, allows for cheap and fast tests of the viability of a new system. In fact, the system need not as yet physically exist. In any modeling adventure, the scientist must make decisions about what fundamental factors are connected and in what way. In cognitive modeling, the endeavor is all the more challenging because many of the fundamentals are unobservable and many of them are only partially understood.

In Project Ernestine, there were a few goals, simple OPS, and straightforward methods. Selection rules were totally absent. In hindsight, we can now look back at telephone operators and argue that the relatively simple nature of the task allowed the researchers to develop a model that could fit on 94 inches of paper. In environments with more goals, OPS, methods, and selection rules, the complexity of the model would naturally increase. However, the increase in complexity of the model is dramatically less than the increase in the complexity of the phenomenon or system that model can explain.

Human factors researchers have developed a variety of modeling techniques and have accumulated a variety of empirical facts that give substance to the constituent ideas that interact within the model.

Today, as was the case 20 years ago, the results of a model can be a voice crying in the wilderness, an unheeded warning about what the future may hold. The value provided by the Project Ernestine team, even after it was initially ignored, was to be able to explain why the new system didn't work. This ability of models to explain and predict makes an invaluable addition to the human factors professional's toolbox. To sponsors astute enough to pay attention, knowing the cost of a cognitive-technical modification in terms of workload and expense is priceless.

Suggested readings

Gray, W. D., & Boehm-Davis, D. A. (2000). Milliseconds matter: An introduction to microstrategies and to their use in describing and predicting interactive behavior. *Journal of Experimental Psychology: Applied, 6,* 322–335.

Gray, W. D., & Fu, W.-T. (2004). Soft constraints in interactive behavior: The case of ignoring perfect knowledge in-the-world for imperfect knowledge in-the-head. *Cognitive Science, 28*, 359–382.

Gray, W. D., John, B. E., & Atwood, M. E. (1992). The précis of Project Ernestine or an overview of a validation of GOMS. In P. Bauersfeld, J. Bennett, & G. Lynch (Eds.), *ACM CHI'92 Conference on Human Factors in Computing Systems* (pp. 307–312). New York: ACM Press.

Gray, W. D., John, B. E., & Atwood, M. E. (1993). Project Ernestine: Validating a GOMS analysis for predicting and explaining real-world performance. *Human-Computer Interaction, 8*(3), 237–309.

John, B., Vera, A., Matessa, M., Freed, M., & Remington, R. (2002). *Automating CPM-GOMS.* Proceedings of conference on Human Factors in Computing Systems (pp 147–154). ACM, New York, NY.

Polk, T. A., & Seifer, C. M. (2002). *Cognitive Modeling.* Cambridge, MA: MIT Press.

Chapter six

"You guys better take good care of me"

Jeanette Liska left her Texas home in October 1990 and headed for the nearby hospital where they were going to repair the hernia that X-rays had revealed a few days before. She wasn't especially nervous about the surgery, although memories of some postoperative nausea from the last time played in the back of her mind. She had spoken to her anesthesiologist about it, and now she even looked forward to "the short rest."

The anesthesiologist's job was to deliver drugs to Jeanette to create a recoverable state of unconsciousness. More precisely, his job was to remove sensation. If the dosage of the administered anesthetic is too high, it will produce cardiac arrest. Even if it does not kill the patient, too high a dose can produce unpleasant, sometimes serious, postoperative consequences for the patient. In addition, anesthesiologists speak of "the walk of shame," when the anesthesiologist follows the gurney of a still anesthetized patient to post-op. So the anesthesiologist tries to administer drugs within a range because either over- or underdosing can be problematic. As Jeanette was about to experience, underdosing can result in terror, panic, and pain.

Jeanette received an injection before she was wheeled toward the operating room. By the time she arrived, she already felt as if she were floating in a warm, relaxing bath. "You guys better take good care of me," Jeanette said right before she was put under for the surgery. Jeanette was in the first stage of general anesthesia—induction. During induction, the anesthesiologist usually follows a standard procedure to achieve the initial state. Today, induction is often achieved by delivery of a bolus dose of drugs intravenously, but it can also be achieved by an inhalation agent.

Ether was the first inhalation agent used in surgery, in 1842, by a 29-year-old rural doctor, Crawford Long, who used sulfuric ether to remove a tumor from the neck of young Jimmy Venerable. Eventually, ether came into widespread use. An interesting exception was childbirth. Women were not given anesthesia both because of concern for the fetus and because it was felt the pain would strengthen the mother–child bond. Prior to ether, patients were either intoxicated or simply expected to put up with the pain. They might be given a belt or a bullet to bite on until the surgery was over or until they passed out from the pain. "Bite the bullet" has become an idiom meaning to face a painful situation, bravely and stoically. The film *Master and Commander* has scenes of several surgeries by a ship's physician during the early 1800s, one in which a young midshipman's arm is amputated while he bites down hard on a cloth-wrapped stick.

Modern inhalation anesthetics are sometimes called *volatile anesthetics* because their natural state is liquid, but they are administered to the patient as a gas. Volatile anesthetics are easy to monitor, in part, because the physics of gases is simpler than the physics of intravenous agents. The anesthesiologist could monitor the gas external to the patient and know that the saturation inside the patient, at the patient's brain to be specific, was the same. Unfortunately, inhalation anesthetics have their problems. They can be stressful to the respiratory system. Some of their pungency is strong enough to make them poor choices for inducing unconsciousness, although they can be used once the patient is under (i.e., during the maintenance phase). Inhalation anesthetics can also cause nausea, vomiting, anemia, and other postoperative problems.

Increasingly since the 1930s, the anesthesiologist relies more heavily on intravenous drugs, although surgeries often use a combination of inhalation, intravenous, and even slow-acting drugs administered preoperatively, like the injection Jeanette received on her way to the operating room. The first record of intravenous anesthesia was 1656, when Christopher Wren administered opium to his dog. Anesthetic agents were first delivered intravenously for human surgery in the mid-1930s.

Jeanette was now in the maintenance phase. This phase is the most complicated, longest phase and ends when the patient is reawakened during the final, emergence, phase. The anesthesiologist delivers drugs, monitors vital signs, and watches the patient's response. He or she monitors proxy variables like heart rate, blood pressure, cardiac output (blood flow through the heart), and blood volume, which give indirect reports on the patient's state. Some signs indicate the dose is too high, whereas some signs indicate that the dose is too low. Tachycardia (rapid heartbeat), hypertension, and actual movement are all signs that the dose may be too low. The anesthesiologist can also monitor brain waves (electroencephalogram). The anesthesiologist uses pen and paper to record significant events and actions in the anesthesia record, which can then be reviewed later in the surgery to give a history of what happened.

Anesthesiologists also have, from years of experience, mental models of the pharmacokinetic properties of the drugs—relationships between doses and concentrations, and the pharmacodynamic properties of the drugs—relationships between concentrations and effects. With such knowledge, an anesthesiologist can project from an administered dose to the particular concentration that will result in 1 minute and then from that concentration to the projected sedation effect. If I give her more sedative, when will it have its effect? How will it interact with the analgesic I gave her a minute ago?

The goals, targets, predictions, and mental models must, of course, be adjusted based on the patient, the surgery, and the anesthetic record of progress thus far. Experienced anesthesiologists have acquired exactly these types of skills. In reality, although the pharmacokinetic and pharmacodynamic properties of major anesthetics are known, no doctor is capable of running those models in his or her head with complete accuracy. Gaps in any doctor's knowledge, together with the multiplicity of patient differences (e.g., age, weight) and the dynamic nature of a changing operation, add to the complexity Further, the complexity of the models makes it unlikely that any doctor, regardless of skill level, could run the mental models precisely and without error.

Jeanette heard people talking and thought how little time the procedure took. She heard her anesthesiologist comment on her attractive breasts, the surgeon apologize for being late, and the nurses laughing. It wasn't until she tried to swallow and couldn't, until she tried to open her eyes and couldn't, that she realized they had not even begun.

The anesthesiologist tries to administer the intravenous anesthetics in a timely way to achieve three effects: sedation, analgesia, and neuromuscular blockade. *Sedation* refers to achieving the state of unconsciousness. *Analgesia* is the elimination of pain. *Neuromuscular blockade* creates the paralysis prerequisite to the surgeon's precise use of a scalpel. This concept of *balanced anesthesia*, where the anesthesiologist views anesthesia as controlling the three components, was introduced after curare (to induce neuromuscular blockade) was used in 1942. It is possible to achieve all three states with one drug, but currently the side effects of such single-drug anesthetics are problematic.

Models of how the drugs work in balanced anesthesia have already been digitized; they are, in fact, used in target-controlled infusion pumps. However, these devices have yet to gain widespread acceptance, in part, because they are very difficult to use, although human factors work has been conducted to understand and improve these devices.

Jeanette couldn't swallow, open her eyes, or tell them she was still awake. She remembered an old Alfred Hitchcock episode where crying kept a paralyzed man out of the vault in the morgue. She tried to cry. She couldn't. Unlike the sedative, the neuromuscular blocking agent that the anesthesiologist delivered was having its intended effect. Jeanette was paralyzed. Only her mind was able to scream, "I am awake! I am awake!" Jeanette heard the surgeon ask for a scalpel.

A third drug, one different from the ones that produce neuromuscular blocking and sedation, was used to achieve analgesia, to control the pain. Was it being delivered correctly like the neuromuscular blocker or incorrectly? Unfortunately, like the sedative, it too entered Jeanette's body too little, too late. Jeanette felt the surgeon's scalpel rip into her. It cut and cauterized. "It felt like a blowtorch," Jeanette would write later. "Molten lead." She felt the flesh peeled back, the stench of burning flesh filled her nostrils. She tried to lose consciousness; death became, in her words, "profoundly attractive."

Jeanette heard a vital-signs monitor beeping. Perhaps it was a rising heart rate, but noting the tachycardia was not sufficient. The surgery continued. Above the panic of being paralyzed yet awake, above the pain of the cauterizing surgeon's knife, above it all Jeanette heard the surgeon speak: "Well, I'll be damned. It's not a hernia. It's just some fatty tissue. All that for nothing."

Intraoperative awareness

Being awake while under general anesthesia is a rare event. It is estimated to occur only once or twice in 1,000 operations, although some suggest that the incidence rate is as high as 1 in 100. Once or twice out of 1,000 is rare: It is about the rate with which sailboats are struck by lightning or women give birth to babies with Down's syndrome. Nevertheless, if 1 of every 1,000 planes crashed, few of us would be brave enough to fly. With 50,000 surgeries a day in the United States, 1 of 1,000 means that 50 people each day will have awareness of their surgery.

A number of factors seems to predict the likelihood of recall under general anesthesia. One source of such information is the database of the closed claims project that comprises insurance companies' closed malpractice cases. From that database, we learn that the gender of the patient is a factor that influences the probability of being awake under general anesthesia. Using closed cases, women (77%) are considerably more likely than men (23%) to remember events from the surgery. This is not merely because women undergo more surgeries or file more lawsuits. In other anesthesiology malpractice cases, women do outnumber men, but only by 18 percentage points, not 54. Women recalling surgery is significantly higher than would be expected from the baseline of anesthesiology cases. Other risk factors include age (under 60 increases the likelihood), anesthetic (no volatile increases the likelihood), and type of procedure, with obstetric/gynecological and cardiac procedures, which require light levels of anesthesia, showing higher rates of awareness. Finally, the awareness experience usually occurs during the maintenance phase of surgery perhaps because the induction phase follows standard, scripted procedures.

The surgery was over, but the nightmare was not. None of the nurses believed Jeanette, thinking she was just remembering dreams. The anesthesiologist didn't believe her until she confronted him with his comments about her body. The surgeon claimed to have repaired a hernia, but relented when confronted with Jeanette's knowledge that the surgery had been unnecessary. Jeanette suffered from post-traumatic stress disorder. Details of her recovery and additional details of the surgery are in her book, *Silenced Screams*. The details of the surgery reported here were based on that book.

Jeanette is not alone. Linda Hinchliffe was awake during a caesarean and listened while doctors discussed preventing the loss of her baby in fetal distress. Carol Weihrer was paralyzed and did not feel pain. She did, however, feel the intense pressure required to remove an eyeball from its socket as surgeons spent 5 hours removing her eye. To this day, she is unable to sleep lying down because of the flashbacks and feelings of helplessness. Both Liska and Weihrer have become advocates for patient safety and the elimination of awareness during surgery.

Cases of awareness are not exclusively the result of substandard care, although certainly many of the malpractice cases are such. However, even in the delivery of substandard care, it is important to determine the cause of these failures in care standards. Rarely is it a lack of concern or attention by the anesthesiologist (only three claims in the closed claims analyses discussed earlier cited these types of factors). About one third of the malpractice cases indicated standard levels of care.

A human factors solution

Frank Drews, of the University of Utah, was watching another surgery. He had traveled hundreds of miles from Utah to take advantage of the opportunity. He was in the operating room that day as part of a team of medical and human factors professionals trying to improve the delivery of anesthetics to patients during surgery. Drews was the human factors lead on a multiple-year effort to understand how anesthesiologists do their job, what they know, and how they use their tools. They were also about to develop a display that would reduce the probability that someone would relive Jeanette's experience.

Drews noticed the patient moved around when the surgeon began to cut the aorta. Open-heart surgery, Drews knew, was one of the operations most likely to exhibit signs of awareness during surgery. The anesthesiologist made adjustments. The patient quieted once again. At an earlier surgery, the anesthesiologist had been really absorbed—talking to the surgeon, to the nurses, and so on. It took him quite a while to realize that the saturated oxygen (SAT), the oxygen carried by hemoglobin in the blood, had dropped to 90%. Normally, the SAT should have been 98% or higher. He looked into the problem, but couldn't find the cause. Over the next 10 minutes, the SAT dropped to 70%, which is a dangerously low level. Finally, the problem was located: The ventilator tube was disconnected. What happened to the patient during this time, Drews wondered? What was the lasting effect of this lack of oxygen? When Drews entered those operating rooms, it was about the science of improving health care. By the time he left, "it was much more about individual patients," Drews would say later. "The phrase 'improving health care' became 'this would not have happened with better technology.' "

In October 1999, almost 9 years to the day after Jeannette suffered on the operating table in Texas, Drews met Dwayne Westenskow on the University of Utah campus. Westenskow, a biomedical engineer in the Department of Anesthesiology, had just received a grant from the National Institutes of Health. Westenskow had, for years, been working on advances in biomedical technologies. It had only been recently that he had the key realization. "We had gone about as far as we could. Further advances required understanding how anesthesiologists used the technology, how they used the alarms. It required cognitive human factors." It was this insight that brought Westenskow and Drews together.

Where does one start so that patients can avoid the nightmare of Jeanette's painful surgery or Carol's panic-filled 5 hours? Like most complex cognitive activities that modern operators are asked to perform, a failure can have many sources. In fact, it is rare to find a major failure due to a single source. There are, of course, very rare cases when the anesthesiologist fails to engage equipment correctly (e.g., leaky vaporizer) or at all (vaporizer off), but more generally the anesthesiologist is not delivering the optimal dosage. Perhaps the clinician failed to adjust for obesity or age; perhaps because most studies have been on men, applying that knowledge to women is inappropriate; perhaps she missed the clues from the patient or the vital sign monitors; or perhaps she miscalculated the effect of the size of the dose, its time course, or its interaction with the other drugs.

We know that the anesthesiologists must monitor signs from the patient. Missing a sign, like the tachycardia in Jeanette's case, could have contributed to the underdosing. Of course some signs, like the patient moving on the table, as Drews witnessed, are easy to see and interpret. Thus, paralysis of the patient is ensured given the easily interpreted movement cues; such is not the case for sedation and analgesia. Based on these signs, the anesthesiologist adjusts the delivery of the drug, taking into account that it will take some time for the drug to work, that not all of it will reach the site, and so on.

Thus, a critical aspect of the anesthesiologist's job is timing the delivery and predicting when optimal concentrations will be reached. If the anesthesiologist does not know the time course of a drug or predicts it incorrectly, the patient's anesthesia can be suboptimal. But, the Utah team reasoned, if that were the case, then it would mean anesthesiologists needed help with their mental model. But do experienced anesthesiologists really need that help?

A student on the Utah team, Paul Picciano, asked exactly that question. He studied resident and attending anesthesiologists, some of whom had over 20 years of clinical practice. The clinicians answered questions about drug delivery. For example, "for a 72-kg, 35-year-old male patient, what bolus dose of fentanyl is required for apnea (ED50)?" The responses ranged from 120 to 1,000 ug. Thus, some the participants suggested giving a dose eight times the dose suggested by others. This kind of variability among experts points both to the complexity of the task and the gaps in knowledge that even experts can possess.

So, anesthesiologists' knowledge was not perfect. That meant that the right technology might be able to serve a valuable function in the operating room. But what would that technology be like?

A critical analogy

Drews walked to one of the team's weekly meetings, with a brisker step than was usual for this time of the day. But today was different. Drews knew the answer. He knew how to conceptualize the anesthesiologist's job, and he knew that once that was understood the team could design the display. In fact, if he were right, designing the display wouldn't be all that difficult.

What Drews convinced the team of that day was essentially that anesthesiology could be better understood if they thought of the anesthesiologist as a pilot. Controlling an airplane requires the pilot to track various aspects of the aircraft, making adjustments in heading, thrust, and altitude at the right time, monitor the results, and readjust. Fighter pilots who do this well are said to have "the right stuff." Anesthesiologists track not heading but sedation, not thrust but analgesia, and not altitude but neuromuscular blockade. Anesthesiologists with the right technology in their "cockpit" could have the right stuff as well.

Much of what a pilot does is monitor events and track responses to those events. Tracking is a common lab task. Tracking tasks are important components of hundreds of industrial tasks, from flying jets to monitoring them as they cross a radar screen. In a tracking task, an operator must move a control point—say, a mouse—so that it tracks a target moving across the screen. The Utah team embraced Drews' analogy. Anesthesiology was a tracking task. It was not a perceptual one with a target moving in front of the person, but a cognitive one where the operator had to track the delivery of three drugs, increasing and decreasing the right drug at the right time. What if, instead of all of this tracking going on cognitively, it was presented to the anesthesiologist in some perceptual form? What if the display gave estimates of concentrations a short time in the future? The team realized that models of pharmacokinetics and pharmacodynamics were already available. Remember the infusion pumps? Perhaps these models could be used as the foundation of a display that presented the information about the drugs in a form that the anesthesiologist could digest rapidly. If the clinician monitored information about the drugs, even predicted information, tracking should be better than when monitoring proxy variables like heart rate and cardiac flow.

Stan's operation and the Utah monitor

This insight led to the design of the Utah drug display. Using models of pharmacokinetics and pharmacodynamics, the monitor presents a continuous display of predicted effect site concentrations and drug effects. To obtain the display requirements needed to design the monitor, the team interviewed a number of expert anesthesiologists. After the interviews, it was clear that the display should show three things: (a) drug dosing for current and past drug administrations, (b) predicted pharmacokinetic concentrations, and (c) combined effects on levels of sedation, analgesia, and neuromuscular blockade. After using iterative design, usability testing, and rapid prototyping techniques, standard weapons in the human factors professional's arsenal, the team produced a display that provides separate graphs for each of the three administered drugs (sedatives on top, analgesics in the middle,

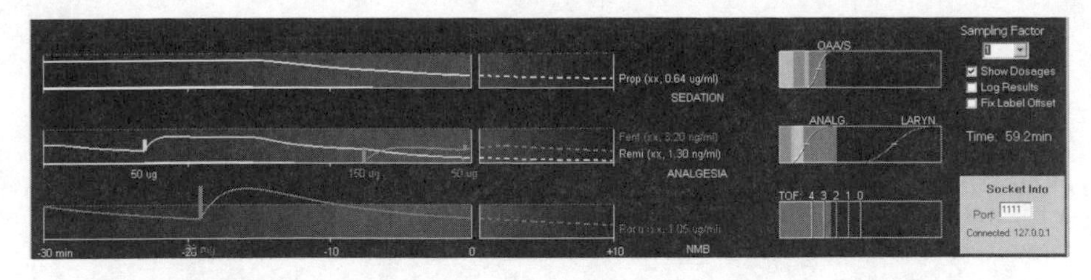

Figure 6.1 The Utah drug display.

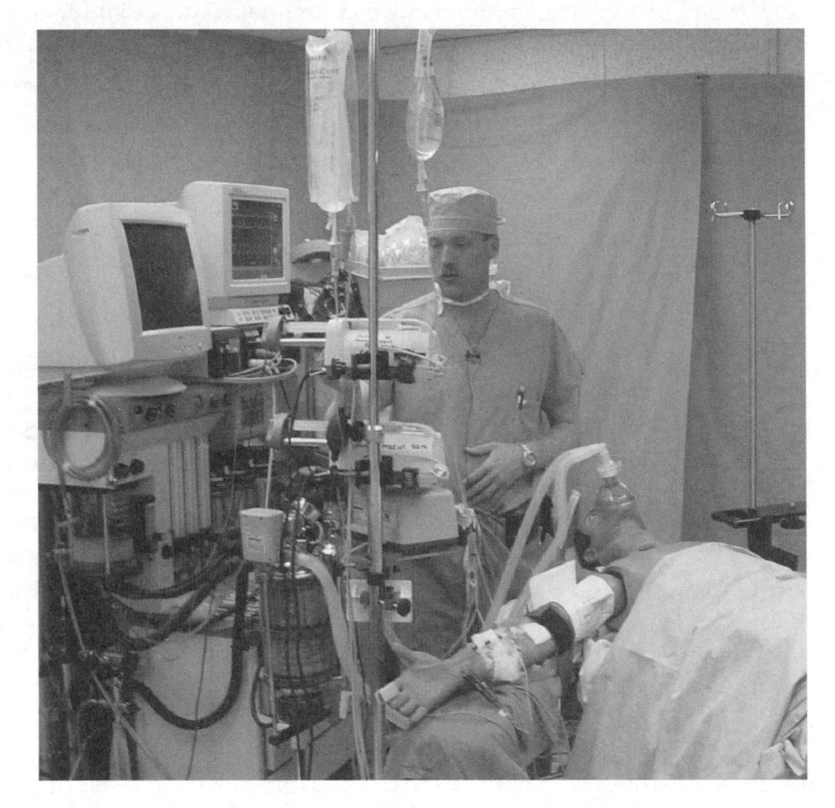

Figure 6.2 Stan, the University of Utah simulator.

and neuromuscular blocking at the bottom). Histograms indicating time of delivery are on the left of each of these graphs in Fig. 6.1. To the right, other graphs show predicted levels of each of the three components of the anesthetic triad, as well as information about drug interactions. The display looked nice, but would it actually help improve how experienced anesthesiologists do their jobs?

The high-fidelity human patient simulator at the University of Utah was a full-sized mannequin capable of blinking, speaking, and breathing. The team referred to him as Stan (see Fig. 6.2). Stan had a heartbeat and a pulse and simulates reactions to a variety of medical procedures. The Utah team was going to see how Stan reacted to anesthesia delivered by experienced anesthesiologists (chief residents, Utah Med school faculty) who did and did not have access to their display.

It turned out that Stan's heart rate and blood pressure stayed closer to the baseline rates during the surgeries that used the monitor. Stan also experienced less pain when the monitor was used. The Utah team was rightly thrilled. All they hoped for had materialized, but Stan had another surprise or two. It turned out that control of sedation was better with the monitor. Stan opened his eyes sooner and reached 50% respiration sooner with the monitor. Because the monitor allowed for more efficient delivery of anesthesia, the surgeries using a monitor lasted 47 minutes, compared with 54 minutes without a monitor. If this 13% drop in operating time were applied to a case like Carol Weihrer's 5-hour operation, she would have saved almost 1 hour, not to mention that she would most likely have been sedated.

Apparently the display did present the information to the anesthesiologist in a way that is readily useable to synthesize the information, make predictions, and administer the drugs. In fact, the performance of an inexperienced anesthesiologist with the monitor was similar to that of an experienced anesthesiologist without one. So expertise seems only a display away.

The BIS monitor and the concept of workforce resistance

Why not simplify the information further, perhaps to one number? If the number is too high, the patient is aware. In fact, another display called the Bispectral Index (BIS) monitor, developed by Myles, Leslie, and the B-Aware group in Australia, does essentially that. The BIS monitor analyzes components of a patient's electroencephalograph and displays a single number between 0 (deep anesthesia) and 100 (awake) based on those brain-wave analyses. A study by the Australian group using 2,500 patients has shown that use of the BIS monitor can also reduce the chances of awareness during surgery.

The BIS monitor is experiencing resistance in the medical community. The American Society of Anesthesiologists has prepared a brochure on patient awareness. In that brochure, brain-wave monitoring devices are explicitly mentioned: "Brain-wave monitoring devices have not yet been shown to enhance the safety of anesthetics given with already-proven technology and the exercise of sound medical judgment."

Why the resistance? Perhaps the brochure is advocating cautious optimism or is a call for additional scientific testing. In fact, there are human factors questions tied to the transparency of the system: interpreting a single number, overreliance and trust in automation, and what to do when the single number goes wrong. In addition to such scientific concerns, however, there may also be a sociopolitical concern. There may be a feeling that too much of an anesthesiologist's job is being usurped by a machine that simply returns one number. Such resistance has a long history, and there are two sides to the story. Most people have heard the term *Luddite* and know that it has come to signify people irrationally afraid of new technology. But when Ned Ludd organized the revolt against new textile techniques, there were genuine concerns about the quality of the product and the employment of the artisans. Perhaps there are similarities?

If there is any truth to such a sociopolitical factor operating with anesthesiologists, we would expect that anesthesiologists would be more likely to embrace the Utah monitor because it leaves the anesthesiologist as the ultimate decision-making authority. The Utah monitor is more likely to be viewed as an aid to help the anesthesiologist do their jobs. As such, it may avoid the type of resistance that has accompanied the deployment of the BIS. Currently, the Utah monitor has been licensed and scheduled for deployment within the year.

Lessons learned

Changes in drug administration, from inhalation agents to a triad of intravenous agents, warrant changes in the environment in which the anesthesiologist works. Awareness during surgery is a catastrophic event that has been on the rise over the past three decades. Such errors are not due to lack of skill or lack of concern on the part of anesthesiologists. Many incidences of operative awareness cannot be traced to simple errors like vaporizer leaks, and even fewer can be linked to procedural error, like failing to turn on the vaporizer in the first place. The Utah team believes that the maintenance phase of anesthesia is too cognitively complex even for highly skilled, highly motivated, caring clinicians. Human factors researchers have developed aids that should help the anesthesiologist. The Utah monitor presents information to the anesthesiologist that would otherwise be hidden, discerned by the inferences of the anesthesiologist based on complex cognitive calculations. The monitor presents this information in a form that is readily understood by the anesthesiologist—the right information at the right time. The monitor is clearly a tool to the anesthesiologist; the person is still in charge of the procedure, not the machine. As long as the ultimate moral and legal responsibility for creating the optimal anesthetic state rests with the operator, then that operator will demand the right to be the ultimate decision maker.

As anesthesiologists accept the additional monitor as part of their arsenal, information currently hidden from the clinician will be presented that matches the needs of the anesthesiologist pilot, allowing him or her to do the right thing at the right time.

Suggested readings

Drews, F. A., Syroid, N., Agutter, J., Strayer, D. L., & Westenskow, D. R. (2006). Drug delivery as control task: Improving patient safety in anesthesia. *Human Factors, 48*, 85–94.

Institute of Medicine. (2000). *To err is human.* Washington, DC: Author.
Liska, J. M. (2002). *Silenced screams.* Park Ridge, IL: American Association of Nurse Anesthetists (AANA) Publishing.
Syroid, N., Agutter, J., Drews, F., Westenskow, D., Bermudez, J., Albert, R., Strayer, D., Prenzel, H., Loeb, R., & Weinger, M. (2002). Development and evaluation of a graphical anesthesia drug display. *Anesthesiology, 96,* 556–574.

Chapter seven

Too many cooks

> The nuclear plant emergency personnel were responding to a Loss of Outside Power (LOOP) event. Oddly enough, nuclear plants actually buy back their own power to cool the plant. A LOOP event can be caused by something as simple as the breakdown of a $2.00 power grid component. However, if not repaired, this minor breakdown can result in the plant's inability to cool itself, creating circumstances ripe for much greater disaster. It is the emergency response team's job to isolate and diagnose the failure causing the LOOP event and to repair it to mitigate disaster.

Emergency response and the systems that surround it have taken a front seat since the events of September 11 and Hurricane Katrina. At the center of emergency response system is a network of humans who are required to share information and function in a coordinated manner. They are to respond to catastrophic events as well as single-component failures that could lead to catastrophic events. In 9/11 and Katrina situations, the system's weaknesses were revealed, and inadequate response was the result. Unfortunately, these two occasions were not anomalies.

The setting for this success story is the emergency response system of a nuclear power plant. Nuclear power plants each have an emergency response system specific to the plant. The system is evaluated regularly by the U.S. Nuclear Regulatory Commission, and training programs and simulated runs are frequently held. It was the early 1990s and with the Three Mile Island disaster only a decade old, emergencies at nuclear plants were taken quite seriously. In a southeastern U.S. nuclear power plant, emergency responders who were notified of the event through their beepers started gathering, and they kept coming. Even those who were not on duty responded. There was no shortage of emergency responders.

So the responders came, and the LOOP event started to run its course. Without the source of outside power, the plant began to heat up. In the increasingly crowded emergency response center, there was confusion. It was not clear what needed to be done. What was the source of the failure? How should it be repaired? Time was of the essence. Equipment began to fail, and the plant was having extreme difficulty cooling itself. If the plant equipment continued to heat up, severe damage to expensive equipment would result. At this point, the only recourse was to force the plant to shut down, creating a devastating financial loss for the company, but preventing major structural damage and potentially, a much greater disaster. However, shutting a plant down is not a trivial matter. It requires knowledge of steam generators and other sensitive equipment within the plant. Fortunately, among those in the crowd of emergency personnel was a single individual who understood the limitations of some of the most vulnerable, most expensive equipment. He was instrumental in shutting the plant down and preventing a financial catastrophe.

Why did this happen?

Although this plant had previously received high scores from the Nuclear Regulatory Commission (NRC) on responses to emergency drills, with the occurrence of this LOOP incident, it was now under increased scrutiny. This meant that the plant faced the possibility that the NRC would increase its number of drills and formal exercises required, each costing between $250K and $1M and requiring upward of 130 participants. How is it that, despite the overwhelming response from emergency response personnel to the relatively simple LOOP event, a disaster was just barely avoided?

First, it is necessary to understand the organizational structure of the Emergency Response Organization (ERO). In this particular plant, as in many plants around the country, the ERO has four components that operate at three levels: (a) the Control Room is responsible for daily operations of the plant and manages any event until the other levels of the ERO are fully staffed; (b) the Operations Support Center includes operations and maintenance technicians who operate at the "ground level"; they hang out in a place called "the kitchen," work hands-on with the equipment every day, and go into the plant to accomplish specific tasks; (c) the Technical Support Center is a level above the Operations Support Center and Control Room; it contains 80 to 85 personnel, including the heads of the major departments, as well as the plant manager; and (d) the Emergency Operations Facility is a level above the Technical Support Center. The Control Room, the Operations Support Center, and the Technical Support Center were all onsite. At this particular plant, the Emergency Operations Facility—the top dogs in this particular organization—gathers roughly 40 miles away. This is by design to enable them to continue their operation in the event of an emergency that requires evacuation of the plant.

On the day of the LOOP event, the ERO was activated and personnel were notified by beeper. This was not a drill! What could the problem be? An air of uncertainty mixed with trepidation permeated the plant as those on duty made their way to their designated spot. Others who were not on duty also moved quickly to make it to the plant. You might think that "the more the merrier" and that only good could come from putting more personnel on the problem. But in this situation, the adage "too many cooks spoil the broth" may be the better take-home message because, despite the numbers of individuals at the scene, there was a lack of coordination with little knowledge among those who were present about what to do, who was to do it, and when.

Those who were at the plant started to assemble at the ERO. As more personnel surfaced, coordination problems seemed to grow. Who does what? Who talks to whom? There were rules in place dictating who should be involved in the case of an emergency. However, there was little guidance about how they should coordinate with one another. Some of the pieces for successful coordination seemed to be in place, whereas some things just didn't seem very efficient. For instance, there was the extended phone call from the NRC official. During states of emergency, the NRC official assigned to the plant contacts the senior shift supervisor of the control room via telephone and stays on the line as plant personnel mitigate the event. This is the procedure, but it tended to remove one of the most senior (and sometimes most expert) individuals from the team. So there were some specific weaknesses of the ERO that could be highlighted, but what was the real problem? What really went wrong, and how could future events like this be better managed?

A cry for help

Following the LOOP event, the managers of the plant recognized their need for outside help with the ERO and went to their friend, Doug Harrington, a team expert at a company called Team Formation. Harrington, realizing that he too needed help, went to the research organization of Klein Associates. The plant managers were very proactive. Unlike many other teams, they saw that they could improve their process, and they wanted some experienced "eyes" to observe them and help them get better.

Things were moving at an accelerated pace. Harrington was called and asked whether he could observe an upcoming drill. Harrington called Klein Associates, who sent Dave Klinger and Tom Miller to attend a drill at the plant only a few days later. The drill, like most drills, involved the simulation of multiple events. To name a few, there was a fire in a junction box as well as the all-too-familiar LOOP event. The observation team videotaped the event, made observations, and interviewed various personnel.

Although the observations and interviews turned out to be extremely valuable, the same cannot be said for the videotape. There was just too much movement of personnel, too much communication taking place over headsets and the computer network, and the role and function of team members was not readily apparent. The ERO—even during a drill—was not the idyllic control room that one might imagine or that has been depicted on the big screen, in which people are all seated at individual consoles participating knowingly in seemingly choreographed discourse with little extraneous noise or interruption. Just the opposite was the case for this drill and most emergency management drills of which we are aware. Noise, interruption, movement, and confusion cloud the environment, making it challenging for operators and observers of operators alike.

As an aside, this shortcoming of video records has been acknowledged by cognitive engineers, and significant effort has been devoted to tools for annotating and analyzing of video. Still it is difficult to capture the larger context of work in a video record. Observations, like the lens of the video camera, can also be focused, but attention can be dynamically and intelligently directed by the human observer based on the questions or hypotheses at hand.

How would one observe in such chaos? Cognitive engineers often focus on particular aspects of the scene that are theoretically or practically the most interesting. For example, the Klein team's observations and interviews were conducted within a conceptual framework for team decision making developed at Klein Associates called the Advanced Team Decision Making (ATDM) model. Actually, ATDM is not so much a formal model like the GOMS model presented in chapter 5, but is rather a summary of important team activities derived on the basis of previous Klein Associates' observations of several types of teams. It identified both individual- and team-level behaviors, such as adjusting, detecting gaps and ambiguity, and defining roles and functions that in the past had discriminated effective from ineffective teams. Thus, the observations and interview questions of the Klein team were guided by this theoretical framework. The framework served as the mechanism for focusing attention during observations.

This is only a drill

The ERO drill lasted between 3 and 4 hours and involved 80 to 85 people in the Technical Support Center alone. Although some communication among emergency personnel was done over the local computer network, there was plenty of voice communication using both land-line phones and headsets with radios. Face-to-face voice communications, obviously, only helped to increase the noise level in the room. Needless to say, it was a loud place.

Klinger and Miller made observations of the Technical Support Center because they were told that this is where the decisions are made. This is where the plant manager resides, as well as the heads of all major divisions. Also, the Technical Support Center is where the media get their information, where evacuation plans are developed, and where critical decisions regarding equipment, personnel, and resources are made. In other words, this was the heart of the decision making in the ERO. Because decision making is the focus of the ATDM model, the Technical Support Center seemed like a good place to start. So Klinger and his team set out to identify decisions, decision makers, and crucial issues in the decision-making process. In short, they set out to expose the decision-making expertise resident in the Technical Support Center (i.e., What decisions are made? Who makes them? How do they make them?), and with 80 to 85 people as possibilities, this was like searching for a needle in a haystack.

Through the initial observations of the Technical Support Center, five key decision makers were identified: (a) emergency director, (b) emergency planner, (c) director of operations, (d) director of maintenance, and (e) director of radiation protection. Interestingly, in interviews with each of these key players, the director of radiation protection, whose job it is to monitor releases and recommend action (e.g., identify areas of the plant that are off limits and develop evacuation recommendations), proclaimed that there was no need for him to be in the Technical Support Center at all. He felt that all his communications were with individuals in the Operations Support Center. Interestingly, the other four key decision makers felt they needed the radiation protection manager there. He was their security blanket—one to whom they had grown accustomed.

One aspect of the Technical Support Center that was observed was the communications flow. On the surface, it seemed overly complicated and, in many cases, just plain inefficient. For instance, the directors of maintenance and operations sat side by side, yet they seldom spoke directly. Instead their communications went through the Operations Support Center.

Klinger observed that in one instance during the drill, the Control Room requested that an operator be sent to close a valve, which set off the chain of communications shown in Fig. 7.1. In response to the Control Room's request to maintenance (Communication #1), the director of maintenance first called the "kitchen," where the operators are located within the Operations Support Center (Communication #2). However, for this request to be processed, it first had to go back to the Technical Support Center to the director of operations (Communication #3), coincidentally seated next to the director of maintenance who first received the call, and the radiation protection manager (Communication #4). These two discussed the situation (Communication #5), made recommendations, and provided specific information relevant to the tasking back to the kitchen (Communication #6). However, the kitchen first communicated their plan to the director of maintenance in the Technical Support Center and got information on assigning personnel (Communication #7) because he was the one who knew the right people for the job and could ultimately send the team (Communication #8). As a result of all of these links in the information chain, it took over 5 minutes to send a team out (an average of 37 seconds per link).

One wonders how communication systems like these evolve. Some of it is procedural—checks and balances to prevent error—and some of it may be the result of patches to an existing communication system put in place like duct tape to fix a previously discovered system weakness (e.g., Joe Smith made a bad call during the last drill and so he needs some oversight). Other links in this chain may evolve over time because people learn to work around the faulty or incomplete procedures or simply because people are confused about the correct procedures. But in many organizations, it is just this process that becomes the bottleneck in decision making. Nowhere is the impact of inefficient coordination more clear than time-sensitive decision making, such as that required for emergency management (e.g., the response to Hurricane Katrina).

Another key set of observations during the initial drill was the response to a simulated fire in the utility room. Notification of the fire was received

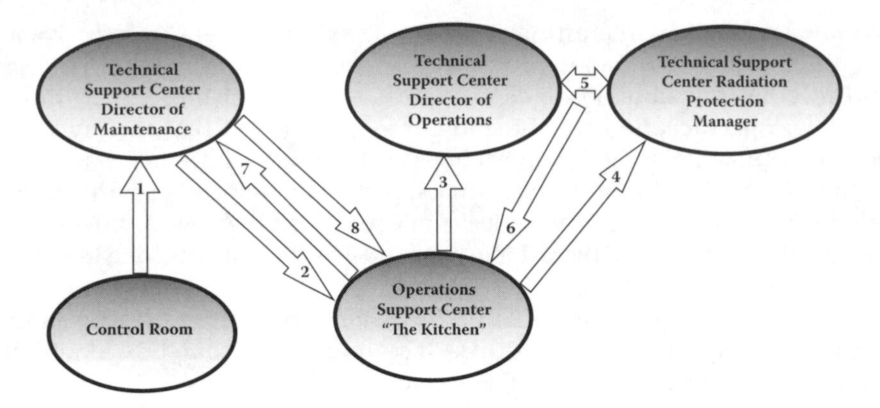

Figure 7.1 Communications required for control room to request an operator to close a valve.

at 12:05. At 12:15, the engineering group was tasked by the plant manager to determine the implications to the plant of a junction box in the utility room becoming disabled. Then, between 12:15 and 12:25, the simulated LOOP event occurred. Klinger and his team observed the engineers working extremely hard in an adjacent room on the junction box problem. At 12:20, unbeknownst to the engineering group, the fire was extinguished. During a time-out discussion at 1:30, the lead engineer proudly prepared to report his team's findings to the Technical Support Center only to be told that the fire had been out for an hour—5 minutes after the engineers had been tasked with the problem. Klinger and Miller watched the engineer as the intense emotional energy that had built up during the problem-solving exercise suddenly drained from his face. He and other members of his team were valuable assets to the plant, and their time was wasted, their expertise unused. They could have been helping out with the LOOP event and resulting leaks and steam generator problems instead. It was apparent that the right people were not at the right places at the right time.

Another observation of inefficient communication centered on the emergency director. At times there was a lengthy line of people waiting to talk to the emergency director, an apparent bottleneck in the system. In what seemed like an attempt to "duct tape" the damaged system, the staff member gathered yellow stickies from the people waiting for 20 minutes or longer. The stickies were to be used in case there was anything important to tell the emergency director.

These are just a few examples of the poor coordination and inefficient communication flow that were observed that day across the entire ERO. Klinger generally observed that many people in the ERO simply served as information conduits with little value added. Sometimes these extra links seemed to increase the possibility for miscommunication. They also noticed that roles and functions of individuals were not clear, which results in the "it's not my job" syndrome. They reasoned that one possible solution to this problem was to replace the practice of having people randomly placed on a week-by-week rotation for being on call to one that attempts to match expertise with the problem at hand—essentially, getting the right people into the right room at the right time.

A solution is proposed

Through a series of three more drills, the research team collected observational and interview data and poured over and over it with the team decision-making framework in mind to identify potential problem areas. Klinger bounced his ideas off the subject matter experts, and this feedback led to iterations and eventual recommendations.

One problem area that was quickly identified was the issue of teamwork. Although there were 85 or so people involved in the initial emergency response drill, the team was not effective. Most people had an assistant, although the assistants did not necessarily assist. Team performance can be severely compromised when many people in a large group fail to take responsibility or have an unclear idea of their role and how it fits into the larger picture. In this case, increases in staff size with no coherent reason as to why people were added was likely the culprit. In some cases, staff members are added based on the assumption that more is better. At any rate, people needed to know who was responsible for what tasks. In the course of the drills, a decision was made that some of the assistants should be eliminated. If there was no clear reason for why someone needed an assistant, other than "I want one," that assistant was removed from the exercise. Noise levels dramatically improved with the reduction of staff. In addition, Klinger later recommended that roles be reviewed at the beginning of an event in a pre-briefing.

These staff reduction and role clarification changes also impacted communication flow. When people understand roles, it becomes clear who needs to communicate with whom, when, and why. The intent of procedures or actions was often lost in the communication traffic jam, which was further exacerbated by some staff serving not to facilitate information sharing, but merely serving as information conduits or, in some cases, blocks to information sharing.

Other suggestions followed from other observations. For instance, Klinger suggested a change to how the ERO room should be laid out. After moving many of the assistants out completely, space opened up. This provided room to implement a command table. The decision makers were now colocated with a clear view of one another as well as a clear view of a modified information white board. Klinger proposed that people be located next to people to whom they needed to talk. In addition, in concert with the director of radiation protection's own views, his position was relocated to the Operations Support Center.

Another recommendation had to do with the content of this board that they faced. Previously, the board was used to track who had been sent into the plant to conduct certain tasks and the time they had begun those tasks. This did little to provide information regarding goals, plant status, or priorities. The board was changed to provide this type of critical information. The board was now a "situation awareness" board, providing much needed information so that the individuals in the Technical Support Center could do their job and make critical decisions in a timely fashion. *Situation awareness* is a concept that originated in the aviation community with pilots describing the need to attend to a rapidly changing environment. This concept has taken hold in many other task domains and is especially pertinent when decisions have to be made quickly on the basis of complex and dynamic events. It is certainly pertinent to the emergency response of a nuclear plant. The information on the new situation awareness board was based on the observed needs in the decision-making process.

Another interesting development impacted situation awareness. During emergency events, the ERO would schedule "time outs." During a time out, each of the major departments (representing more than the five identified decision makers) would stand up and report on the status of their departments. These reports often included historical data ("This happened at this time") and questions ("This happened and we don't know why"). These time outs did little more than provide the ERO personnel with a chance to take a short break, get a bite to eat, and relax a bit. After watching this play out a couple of times, Klinger asked them to include in these reports answers to a series of five questions:

"What is the priority in terms of mitigating the event right now?"
"What is the immediate goal of your department?"
"What are you doing to achieve that goal?"
"What will your situation look like in 15 minutes?"
"How many personnel are in the plant?"

The new question set was originally intended as a training tool so that individuals could identify disconnects and improve coordination. But this also served as a tool to calibrate situation awareness or the team's understanding of the immediate situation. This procedure was so helpful in the minds of the plant managers that it was later used spontaneously during a small actual event by the control room personnel to calibrate the team's situation awareness. These kinds of simple, but effective, tools empowered the plant managers by giving them the ability to help themselves on an ongoing basis.

One thing to note about these recommendations is that they are not technological. That is, despite a typical knee-jerk reaction by many involved to throw technology at the problem, most of Klinger's solutions did not do this. Instead they involved staff, facility, workspace, and procedural changes.

In an unprecedented result in cognitive engineering history, and to everyone's surprise, shortly before their much anticipated USNRC-evaluated drill, the plant decided to implement nearly all 50 of Klinger's recommendations.

Further, they even implemented some changes without testing the value added. Many of the recommendations had been implemented in previous drills, but certain large ones had not. For example, the room was totally reconfigured to reflect Klinger's recommendations. Wireless headsets were given to all personnel in the Technical Support Center, cameras were placed in the Operations Support Center with video feeds to the Technical Support Center so the decision makers could see what was going on there, and several more personnel cuts were administered (reducing the overall count of individuals in the Technical Support Center to about 35, less than half the original number). The plant was motivated to make all of these changes to avoid continued and additional penalties.

Put to the test

The changes were made, and the drill was attended by the NRC, heightening tension and concern around the plant. Yet the plant did exceptionally well. There were major improvements, and the NRC gave the plant high scores. The plant did so well, in fact, that one individual erroneously assumed that the plant lucked out due to an easy scenario, only to discover that it was actually one of the more challenging ones. As a result, the NRC took the plant off the two-drill-per-year watch list, thereby saving the plant millions of dollars.

Following this success, table-top drills were added that involved only 5 to 10 people who would sit around a table and talk through the scenario, describing their roles and functions. Based on the observed weakness in coordination and the ensuing recommendations, this type of practice was critical, yet much more cost-effective than the full drills involving 130 people. Eventually, this practice became so entrenched that the director of operations began to spontaneously review roles and functions of individuals at the beginning of each drill.

Although the bottom line here is the improvement in the drill performance of the ERO team, compared with the poor performance associated with the real LOOP event, other metrics of success are also available. The five decision-making experts were asked to rate the team's performance over the course of the five drills, including the fifth NRC evaluated drill, on a five-point scale. Each successive drill included additional interventions in which Klinger's recommendations were implemented. The expert ratings, which are overlaid on the figure indicating numbers of staff per drill, indicate that there was improvement (i.e., higher ratings) over the course of the five drills. Interestingly, there was a concomitant decrease over drills in the number of participating staff members that went from 70 to 35. This drives home the point that bigger teams are not necessarily better teams. Although these data should be viewed as preliminary, they reflect the scores received as well as the plant's opinion of their own performance (see Fig. 7.2). The research team heard anecdotal evidence following each drill that "things seemed better,"

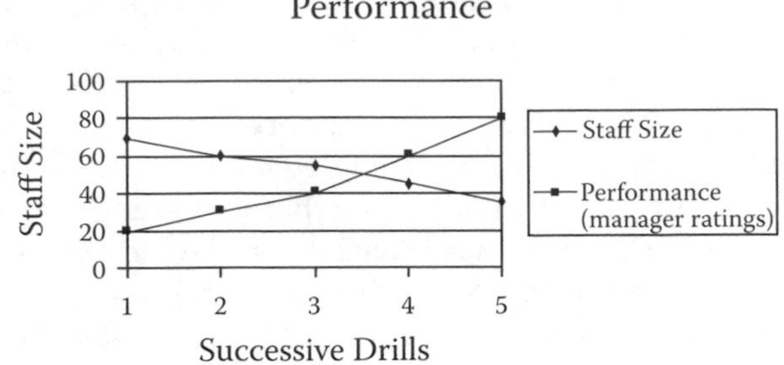

Figure 7.2 Performance improves and staff size decreases over five drills.

"I liked how much quieter the room was," and "I thought I needed an assistant only to find that I spent more time telling them what was happening than I did doing my job."

The success of this project led to another one on control room decision making and training. Many of the same findings and recommendations were applicable to this setting. For example, the situation awareness debriefs were implemented in the control room during both actual and training events.

Lessons learned

There were several lessons learned that can apply to many large-scale, team data-collection efforts and organizational redesign:

- The use of videotape is limited in a large-scale, distributed, team environment. This is particularly true if team members are communicating via headsets and computer networks.
- Be careful of pat answers. When this team asked, "Who are your key decision makers?" they were given a list of five people. As it turns out, those were not the right five. Individuals answered that question based on the organizational chart. They picked the top five people and assumed they were the ones making the critical decisions. Through interviews and observations, the team determined that the list provided by the exercise participants was misleading.
- Too many cooks CAN spoil the broth. The additional assistants and "helpers" simply made the room too noisy, too busy. When assistants are added, they are typically individuals with less experience. Therefore, certain tasks for which they are to assist are simply beyond their level of expertise. These often become mentor/apprentice situations instead of decision maker/assistant situations.
- Technology is not always the answer. Often low-tech fixes can be just what is needed to repair the problem.

- You cannot underestimate the value of skilled observers and their ability to help an organization diagnose and repair its own weaknesses. The patches and duct tape applied to organizational problems reflect an inability to see the bigger picture and the complex interactions in a system of systems. Observing and developing an understanding of some of this complexity is one of the values added by cognitive engineers.

Suggested readings

Durso, F. T., & Gronlund, S. D. (1999). Situation awareness. In F. T. Durso, R. S. Nickerson, R. W. Schvaneveldt, S. T. Dumais, D. S. Lindsay, & M. T. H. Chi (Eds.), *Handbook of applied cognition* (pp. 283–314). New York: Wiley.

Klinger, D. W., & Klein, G. (1999). Emergency response organizations: An accident waiting to happen. *Ergonomics in Design, 7*(3), 20–25.

Militello, L. G., Kyne, M. M., Klein, G., Getchell, K., & Thordsen, M. (1999). A synthesized model of team performance. *International Journal of Cognitive Ergonomics, 3*, 131–158.

Perrow, C. (1999). *Normal accidents: Living with high-risk technologies.* Princeton, NJ: Princeton University Press.

Reason, J. (1990). *Human error.* Cambridge, MA: Cambridge University Press.

Chapter eight

Decisions at sea

On July 3, 1988, at 10:17 a.m. (Iran time), Iran Air Flight 655 took off from Bandar Abbas. Captain Mohsen Rezaian and 289 passengers were departing 27 minutes after the scheduled departure time of 9:50 a.m. It was to be a brief 28-minute flight to Dubai.

At the same time, the USS Vincennes (see Fig. 8.1), a U.S. Navy-guided missile cruiser equipped with an Advanced Electronic Guidance Information System (AEGIS) and advanced tracking radar, sailed nearby in the Strait of Homuz. For Captain William C. Rogers III and his crew, the air was heavy with tension and fear. Two Iranian gunboats, The USS Sides and the USS Elmer Montgomery, were engaging the Vincennes in a battle that had the cruiser swinging around and creating significant noise and confusion. To make matters worse, there had been intelligence reports of predicted attacks on the United States around the 4th of July.

Within the Combat Information Center on board the Vincennes, the atmosphere was similarly stressful, although somewhat removed from the gunboat battle. The technology was new to many on the ship, and the displays were confusing. The room was dark, but bathed in red light to keep eyes accommodated to see the radar screens. Although the ship was swinging around in its battle with the small boats, there was no view from the Combat Information Center of the surrounding area.

At 10:47 a.m., Iran Flight 655 appeared on the radar screen in the Combat Information Center. Vincennes operators identified the commercial aircraft as an Iranian Air Force F-14A Tomcat in an attack run mode, although there was some confusion in the Combat Information Center about whether the airbus was descending or climbing. Attempts to contact it over civilian and military radio channels led to no response. Captain Rogers issued a total of seven warnings from the Vincennes that the plane would be shot down, but still no response. At 10:54 a.m., less than 10 minutes after the plane was detected on the radar, it was given one final warning and shot down with two SM-2 missiles. That day, with decisions that took place in a 7-minute period, 290 individuals, including 66 children, lost their lives as the airbus crashed into the sea.

Sharp-end failures and latent conditions

The downing of the Iranian airbus was the final action in a string of events, many of which were, as James Reason puts it, "sharp-end" human failures and others, the result of "latent conditions" or hidden pathogens of the

Figure 8.1 The USS Vincennes.

system. The ultimate failure leading to the shooting was the misidentification of the Iranian Airbus A300 as an Iranian Air Force F-14. How could such a case of mistaken identity happen? Part of the answer lies in the situational context and the system context or climate surrounding the incident.

Iran and Iraq had been at war for the last 7 years, and in the last year the United States had become involved in numerous skirmishes. Both Iran and Iraq had also attacked U.S. ships, including a 1987 attack by Iraq on the USS Stark, costing the lives of 37 Americans. Then the intelligence reports of predicted attacks on the United States around July 4th gave a sense of urgency to the lingering premonitions of attack. This climate of apprehension, information overload, and the metaphoric "fog of war" seeded expectations of an enemy attack.

The historical context created an atmosphere that produced expectations, not too much different from a lightning strike creating expectations of thunder or the words in a sentence creating the expectations of other words (e.g., "the boy jumped over the white picket ______"). Psychologists call these contextually driven expectations *top–down effects* or *top–down processing*, with the top in this case being knowledge or beliefs that guide perception and reasoning. Captain Rogers and his crew were expecting an attack and perceived information in the environment under the veil of these expectations. But the expectations alone would not have been enough to prompt the actions that happened that day.

In addition to these top–down effects, there were also some other conditions that misguided decision making. When data or information in the environment drive perception, the processes are labeled *bottom–up effects* by psychologists. The top–down and bottom–up labels are interesting in that they imply that perception is driven by processes acting in one direction or another, like a ball in a pin ball machine or a yo-yo. On the contrary, most psychologists see perception as a cycle, like a ball in a roulette wheel, with top–down and bottom–up processes working simultaneously to guide perception. Bottom–up processing of information in the environment generates expectations that, as a model by Neisser posits, provide the guidance for the top–down processing of additional information in the environment in a seemingly endless cycle of perception. Psychologists, including Gary Klein, have similarly described perception as a process of recognizing patterns in the environment, but this recognition process is "primed" or biased by knowledge and expectations. This view is known as "Recognition-Primed Decision Making."

What were some of the bottom–up influences leading to the decision to shoot? First, information was ambiguous regarding the identity of the aircraft. Ambiguous information, by definition, must be resolved by some processing other than processing of the environment. Thus, top–down processing, often generated by the context, takes center stage to influence the interpretation of the bottom-up perception. The radar displays were also confusing. For example, the airplane's location was separated on the display from the airplane's vertical movement. Here again the top–down processes can be called on to help make sense of information that is confusing, and in this case helped to paint the inaccurate picture that Iran Air 655 was descending rather than ascending. Second, the physical environment was constraining. There was confusion in the Combat Information Center, which was dark, but bathed in red light. Third, there was no view from the Combat Information Center of the area surrounding the ship. With little information for bottom–up processing, the top–down processes can dominate. Fourth, there were deficiencies in the skills and abilities of many of the crew members who were often not well versed in computer technology (and thus modern warfare) and seemingly unprepared for decision making under these stressful conditions that demanded speed. This was, in other words, a training deficiency. Top–down, bottom–up—it all had to happen in the course of the 7-minute interval, leaving little time for verifying perceptions and opening the door for misperceptions. Thus, the confluence of the latent conditions, associated with unforgiving technology, poor physical conditions, and limited skill, created a collection of resident pathogens or weaknesses in the system that lay dormant waiting for the precise context that would trigger the tragic action.

The immediate aftermath

The Vincennes incident was the impetus for the Tactical Decision Making Under Stress (TADMUS) program that followed from 1990 to 1999 and involved teams of researchers in Orlando and San Diego. Perhaps it is not surprising that the Vincennes incident triggered intense interest in, and dedicated resources for, finding a solution. Too often, however, the solution is thought to reside in advanced technology—more automation, better battle command technology, or faster ships. The TADMUS program represented a human-centered approach to design, in which system or training development revolves around extending the capabilities of the human operators. How did this come about?

Several events came together to provide impetus for the TADMUS program. First, there was an appreciation for this approach at high levels in the

Navy and government. There were calls for research on decision support, stress, Combat Information Center training, and information displays from, for example, Rear Admiral Fogarty and Admiral Crowe of the U.S. Navy and from the House Armed Services Committee.

In fact, the House Armed Services Committee held congressional hearings on the Vincennes incident in the fall of 1988. Fortunately, the individuals involved in these hearings were selected from among the ranks of notable decision-making and human factors psychologists, including Paul Slovic, Richard Nisbett, and Dick Pew. The panel's goal was to advocate for increased Naval funding for research on decision making under stress. Not surprising, the Navy countered with shipboard success stories in which good decisions were made under stressful conditions. Just as failures and latent conditions came together in sparking the Vincennes incident, so did events come together to support the TADMUS solution. The timing was very good. Converging on the same objective as the panel was support for work on cognition, decision making, and training coming from the Office of Naval Research and from a workshop initiated by the American Psychological Association. Together these events generated the impetus for funding for TADMUS. The program was originally slated for 6 years of support and a list of five decision support and training tasks, but was extended for 3 more years and given an additional task to integrate training and decision support.

The TADMUS program

Who were the TADMUS heroes and heroines? There was a large cast, but widely mentioned by those who were involved in the endeavor was the technical advisory board for TADMUS, which reviewed the program every 6 months for the first 6 years. Apparently this was a pivotal structure with seasoned leaders such as Marty Tolcott and Bill Howell, who, according to many, contributed immensely to the success of the program. The remainder of the technical advisory board was made up primarily of Navy senior scientists and active duty fleet representatives. This mix of scientists and operational Navy people also turned out to be a key factor in the success of the TADMUS program.

Key in the TADMUS organizational structure was the fact that it was distributed over the two U.S. coasts. The team training thrust was headed by NAVAIR (formerly Naval Air Warfare Center Training Systems Division) in Orlando, FL, and the decision support system design thrust was conducted by SPAWAR (formerly Naval Command, Control, and Ocean Surveillance Centers Research, Development, Test, and Evaluation Division) in San Diego.

The independent operation of the East Coast and West Coast groups presented some coordination challenges. Not only were these two groups separated geographically, but also conceptually. The Orlando group tended to be psychologists with industrial/organizational backgrounds, whereas the San Diego group tended to be system designers with some human factors background. There were separate budgets, separate promised deliverables, and different perspectives—a breeding ground for a "mine is better than yours" mentality when it came to research and development. Indeed, there was some political maneuvering as one might expect. Also with such a big organization with radically different perspectives, there were many diverse leads pursued, but little integration. This resulted in some perceived duplication of effort and missed opportunities for interaction. The environment required significant efforts to focus. However, the grand plan was to organize an overarching experiment that would demonstrate an integrated training and design solution. In the long run, this goal, the advisory board, and the unifying drive to make a difference trumped politics and territorial struggles to bring the two groups together in pursuit of a solution.

Overall, in the early days, the TADMUS program was seen to many as a risk. Bill Howell and others were taken by the significant costs and benefits of such an endeavor. That is, the risk was great given that there could be a failure of applied psychological science to produce effective solutions. However, the benefits of success were also great, not only for improving shipboard operations and avoiding future incidents like the Vincennes, but also for demonstrating the value of applied psychology to the field.

The solutions

Throughout the TADMUS program, Navy scientists worked with contractors and the operational community to come up with solutions to the types of problems that surrounded the Vincennes incident. Some of the solutions involved design changes, such as modifications of the AEGIS radar user interface of the weapons control officer workstation (see Fig. 8.2).

Ron Moore, Richard Kelly, Sue Hutchins, and Cap Smith of the San Diego group had primary roles in the development of the Decision Support System (DSS) prototype, a display intervention for problems with dynamic tactical decision making under uncertainty and stress. Some of these relatively minor changes (use of color) have been adopted by Lockheed and are now state of the art in the AEGIS system. Some bigger solutions involved the design of a DSS for command-level decision makers. These systems would take in sensor data and help the decision maker integrate that information to make the best decision.

The DSS system was designed "by the cognitive engineering book" using a three-pronged attack on the problem. First, models and theories of human information processing and decision making were considered, especially in light of the errors that led to the Vincennes shooting. For instance, researchers realized that many of the existing decision theories and debates did not seem relevant to the decisions that were made aboard the Vincennes. Second, the Combat Information Center on board a ship provided an ideal

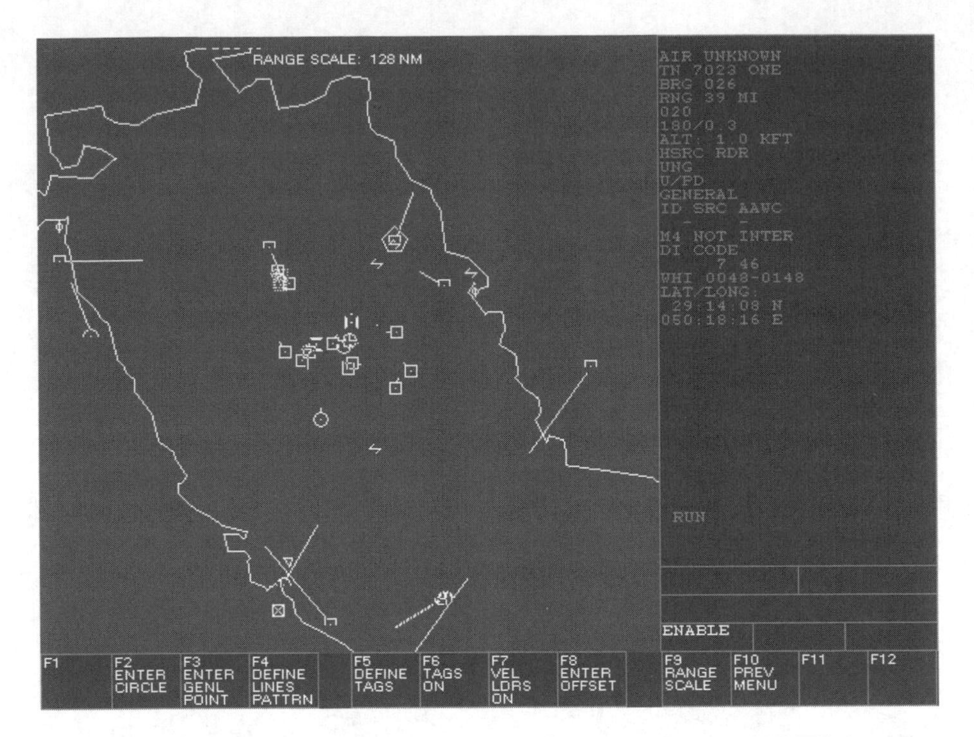

Figure 8.2 Advanced Electronic Guidance Information System (AEGIS) interface circa 1988.

environment for understanding naturalistic decision making (i.e., decision making in the context of complex, real-world tasks), which yielded concepts such as Recognition-Primed Decision Making.

This phase of basic inquiry took several years of the program and was accompanied by several false starts. Indeed, it is often the delay caused by this background research phase that gives engineers and developers the most heartburn when it comes to interfacing with cognitive engineers. However, in this case, despite the false starts, the background phase set the stage for much of the success in the second and third phases.

In the second phase, the specific AEGIS radar task of track identification, including distinguishing friend from foe, was examined in detail to generate a human performance baseline. Cognitive task analysis and naturalistic decision-making approaches (e.g., interviews, observation) were directed at commanding officers in the task environment. A list of errors and cognitive bottlenecks was generated. For instance, operators tended to mix up radar track numbers associated with the different aircraft, or there were problems remembering which actions had been taken on which radar tracks. In addition, information requirements were elicited from operators. It was determined through cognitive task analysis that every operator needed to know about every critical track of interest—the answers to: Who is he? Where is he? Where has he been? Can he shoot me? Can I shoot him? This information, combined with the theory and models in the literature, led to the design of a DSS.

The third step in the three-pronged approach involved iterative design and testing with actual fleet sailors. For instance, three or four versions of each of seven displays were compared for the first version of the DSS (see Fig. 8.3).

The resulting TADMUS DSS displays were a far cry from the original AEGIS radar display—the one used during the Vincennes incident. The San Diego group spent hours agonizing over every square inch of the display, all in an effort to ensure that decision makers had unambiguous and required information that they needed to make critical decisions in stressful environments. The goal was to make cognitive processing of the displays as effective as possible, thereby conserving valuable cognitive resources for other decision making and reasoning. For instance, color was used differently. A monochrome blue background does not adequately distinguish land from sea. Desaturated gray tones were selected instead for the background for several reasons, such as optimizing color contrasts among moving objects and land features.

Another obvious difference apparent in a comparison of Figures. 8.2 and 8.3 is the use of two screens for the TADMUS DSS, on which a plan view (looking down from above) and a profile view (looking from the side) are presented. These two displays, in conjunction with a track history feature involving the dropping of virtual breadcrumbs by objects to show where they have been, made it less effortful for the operator to determine whether an object was climbing or descending.

In a realistic simulation, it was found that the DSS reduced the false alarm rate (e.g., like the 1988 error to shoot down the Iranian airbus) by 44% and increased the hit rate (e.g., correct decisions to shoot down targets) by 21%. In other words, the targets were identified correctly more often, and nontargets were less frequently misidentified as targets.

Many of the DSS concepts are now being used in the development of SPAWAR's "Knowledge Wall" for use by the Strategic Command Center as well as several Navy command centers. However, despite these distinct successes, considering the scope of the TADMUS program, relatively few developments, including the DSS, have been transitioned into the fleet. Also, the transitions have occurred at a slow pace. There are a number of reasons for this disconnect, chief among them being that interface improvements would have to be supported by the developer of the system (e.g., Lockheed) and made in concert with the development of the rest of the shipboard systems, a process that takes 10 to 15 years if the change is introduced at the beginning of a design cycle. Of course, in system-of-systems design, such as that found on a Naval ship, the change introduced would need to be integrated with other existing and new systems. Another explanation is that technologists are reluctant to field their developments too soon. Premature insertion of technology into the operational environment could fail, leading to loss of trust and possibly research funding, not to mention the chance of system failure or even a disaster on the same scale as the Vincennes. These

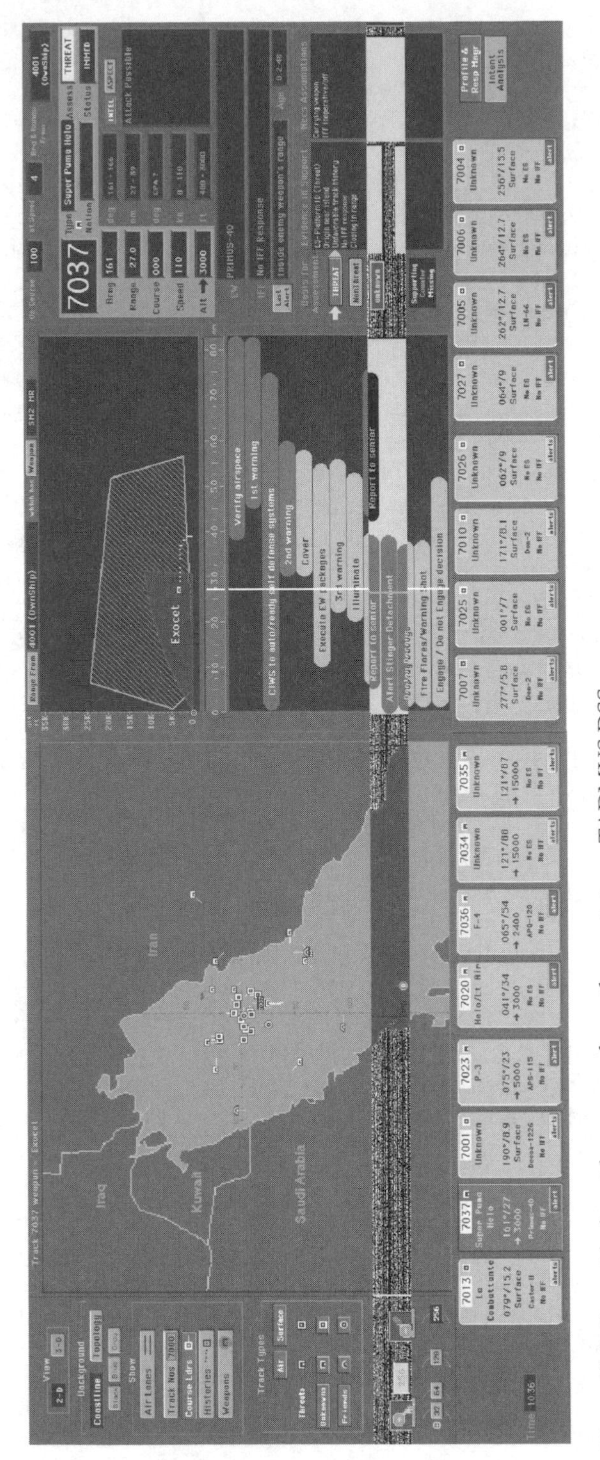

Figure 8.3 Two side-by-side screens from the prototype TADMUS DSS.

challenges, due to size and complexity of the AEGIS system, made it difficult for the San Diego group to have immediate impact.

However, training interventions could be transitioned much more easily, in that they simply required buy-in from those doing the training and often a lead time of only 24 hours to implement a change (contrasted to 10–15 years for design interventions). Training solutions included specific training strategies and methods of measurement. For instance, Kimberly Smith-Jentsch of the Orlando group was instrumental in the development of Team Dimensional Training, a training approach to improve communication and coordination on board the ship. This training targeted group-level decision making that involved knowledge sharing and coordination among a variety of personnel and parallels Crew Resource Management discussed in chapter 4 in the context of airline crews. Team Dimensional Training was introduced to the operational community and fielded in some training programs. It was well received by the fleet, and survey data indicated that the operational community judged it to be a success. There are field data suggesting that the training interventions improved decision-making performance by 40% to 50%. As another indicator of success, this training regime has been adopted by other organizations, including National Aeronautical and Space Administration (NASA) and medical and commercial industries.

The real success

Thus, the TADMUS program was productive in generating a number of specific training and design solutions, although the training solutions were more easily fielded than the design solutions. However, at the same time, a higher level solution to the bigger problem of inadequate human systems integration was unfolding. To the extent that the design and training solutions were successful, the TADMUS program was drawing attention to the benefits of human systems integration, naturalistic decision making, and cognitive engineering. Therefore, the potential was there for a solution to generic problem that would have widespread effects on human-centered design and training even outside of the Navy.

Indeed, attention was drawn to the research. TADMUS research findings were widely disseminated with the program spawning over 200 publications. However, most of this communication was intended for the human factors research community. More was needed to make a broader impact outside of this community. Thus, TADMUS scientists and engineers did more than publish in their peer-reviewed technical journals. They briefed the operational community as well, some of whom were on the TADMUS team. After 6 to 7 years, individuals, including Eduardo Salas, program manager of the Orlando group, began to notice a change in this community. Changes occurred from the shipboard personnel and Captain Ted Hontz, Commanding Officer and head of the AEGIS training center for the USS Princeton, to the Chief of Naval Operations, who got the message that technology is an enabler and that targeting human performance is essential to success. This message was no better exemplified than in the Vincennes incident, which occurred on an AEGIS cruiser with arguably the best battle management technology available at that time in the world. Yet in the end, it was the human decision making, acting in concert with this advanced technology, that failed. This message spread within the operational community and began to change the culture or philosophy of the organization.

Therefore, in many ways, the TADMUS success story is really about a change in an organizational climate. Success came in some products and findings, but mostly in changes to the way people thought about human factors and cognitive engineering. The specific TADMUS operational problems infiltrated much of the human factors research community. Also, the operational community now had a new way of thinking about problems from the user's perspective. In fact, human factors continues to have an important hand in the design of new Navy technologies as a direct result of the TADMUS success and the legacy of the failure that culminated in the Vincennes tragedy.

Lessons learned

The TADMUS program is a success story on both local and organizational levels. The success may be attributed to the perceived enormity of the Vincennes tragedy that triggered the program. Yet the Vincennes incident is certainly not the only high-profile disaster that gained national attention in recent years. Three Mile Island, the Challenger and Columbia accidents, Hurricane Katrina, and 9/11 are not associated with parallel cognitive engineering successes on this level. There are other factors that came together to make TADMUS a success.

When those involved in TADMUS talk about its success five main themes emerge as pivotal (Figure 8.4). First, there was a large and sustained effort dedicated to improving individuals' and teams' decision making in a complex setting, such as Naval Combat Information Centers. The Navy supporters and program managers were as committed to solving this problem as the academicians and contractors who truly wanted to further the science of decision-making training and team performance in this applied setting. The level of funding that TADMUS received was unprecedented for applied behavioral science, but was needed to address such a complex problem.

A second characteristic of the TADMUS program that is partially responsible for its success is the technical advisory board. As mentioned before, this board contained both scientists and operational Navy people. This mix was to ensure that the research coming out of the program was not only scientifically sound, but also had operational value. As Eduardo Salas puts it, the research had to be "scientifically based, but practically relevant." The board helped keep the program on track in achieving its directives through regular meetings and a clear plan with milestones. The board was also instrumental in integrating or unifying TADMUS work across the two coasts.

Factors that Made TADMUS a Success
1. Large and sustained effort/funding
2. Effective technical advisory board
3. Work was communicated broadly
4. Human-centered approach
5. Integration of researchers and operational community

Figure 8.4 Themes pivotal to success of the TADMUS program.

Third, the TADMUS work was communicated through publications, presentations, and demonstrations. The work was communicated broadly, and there was foundational research and applied work that followed on its heels. Simulation was also heavily used as a compromise between sterile laboratory research and difficult-to-generalize field work. Most important, the work and the message about human-centered design was communicated to the broader Navy.

The fourth characteristic of the TADMUS program that is partly responsible for its success is the commitment of those involved to a human-centered approach to design and training. The needs, capabilities, and limitations of the human operators were not only considered, but they were central to the design of technology and training programs. For instance, design of decision support systems drew directly from cognitive findings and theories to develop a system that processes and presents information in a way that supports the decision maker's thought processes. Of course the achievement of this objective hinges on access to the operational decision makers, the fifth and final winning feature of this program.

Last, but likely most crucial for the success of TADMUS, was the intense participation of the Navy training and operational community. There were senior officers involved as well. This was not just business as usual, but a strong connection between the research and the people for which it was intended and their commanding officers. The researchers on the team were committed to operational relevance and to transitioning the results to make a difference for the Navy. Therefore, they made sure that research was validated in experiments with active duty Navy participants both in training units ashore and on board ships. There was a strong desire to understand the Navy tasks and the people who performed them. Therefore, the researchers either immersed themselves in the Navy task or included operational individuals on the research team. Researchers went beyond hand waving about the importance of the user or operational task. They spent significant time in the operational environment. Overall, the TADMUS researchers insisted that a positive relationship develop between the two communities, and it did. Trust and respect developed between the two communities, and both took ownership of the problems, the solutions, and the TADMUS program.

The TADMUS legacy

It is now nearly 20 years since the Vincennes incident and 7 years since the TADMUS program came to an end. But the legacy of TADMUS lives on not only aboard naval ships, but also in the scientific knowledge, training innovations, and decision aids developed under the program. However, most important are the lessons learned from TADMUS regarding the benefits that come from the marriage of cognitive engineering and operational personnel and the risks we face when engineering proceeds without it.

Suggested readings

Cannon-Bowers, J. A., & Salas, E. (1998). *Making decisions under stress: Implications for individual and team training.* Washington, DC: American Psychological Association.

Crandall, B., Klein, G., & Hoffman, R. R. (2006). *Working minds: A practitioner's guide to cognitive task analysis.* Bradford Books.

Fogarty, W. M. (1988). Formal investigation into the circumstances surrounding the downing of a commercial airliner by the U.S.S. Vincennes (CG 49) on 3 July 1988. Unclassified Letter Ser. 1320 of 28 July 1988, to Commander in Chief, U.S. Central Command.

Klein, G. (1989). Recognition-primed decisions. In W. Rouse (Ed.), *Advances in man-machine systems research* (Vol. 5, pp. 47–92). Greenwich, CT: JAI Press.

Neisser U. (1967). *Cognitive psychology.* New York: Appleton-Century-Crofts.

Reason, J. (1997). *Managing the risks of organizational accidents.* Burlington, VT: Ashgate.

Chapter nine

In their own words: the cognitive engineers speak from the trenches

The stories in this book give a perspective on the particular success story in cognitive engineering and tell us about the process leading to each contribution. The current chapter provides additional insights into the process of doing cognitive engineering by exploring the thinking and opinions of the cognitive engineers who contributed to each success.

In late spring 2006, we asked representatives from each story to answer some questions (Bill Howell was asked to speak to the Vincennes story, rather than his general commentary). Some of the questions were intended to elicit insights about the particular endeavor, whereas other queries were targeted at the field more generally. We occasionally edited the comments for length, but what follows is in the words of the scientists.

As you read the answers juxtaposed to each other, you'll see that at times there is a surprising amount of agreement, whereas at other times each contributor comes to the question from a unique perspective.

We have organized the questions and answers into three sections. The first, "Doing Cognitive Engineering," focuses on the projects described in this collection of stories and the scientific underpinnings that went into their solutions. The second, "Becoming a Cognitive Engineer," is a collection of questions that is directed at the student; the questions explore the training and preparation needed to enter this exciting field. The final section, "Cognitive Engineering in Society," wrestles with the challenges researchers face when they try to offer new solutions to well-established problems.

Doing cognitive engineering

In the following section, the reader will notice that, although researchers are in general agreement about how to do cognitive engineering, like scientists in any field, they bring their unique perspective, even to defining what they do. Overall, it seems clear that cognitive engineering involves applied work in often very complex real-world environments. The complexity of the problem was mentioned either directly or indirectly as an intriguing part of each researcher's project. In each case, the complexity was respected, but cognitive engineers use methods to reveal fundamental aspects of the complex situation. To solve the problems, the researcher must be prepared to bring to bear an arsenal of methods from the tried and true to the innovative or high-tech ones; the notion of using multiple methods is mentioned by several of our researchers. Note also that the scientists see success as multidimensional, ranging from financial improvement to personal satisfaction, as well as solving the specific problem. Interestingly, even after achieving success, they often leave a project with the desire to do more.

How do you think of the difference between basic and applied research?

KLINGER (Too Many Cooks): As little as possible.

BALL (Not Too Old to Drive): Basic research in my field is based on theory—primarily testing hypotheses about underlying mechanisms in cognitive function. Applied research should be the next step—taking what has been learned from more basic studies and attempting to evaluate it in applied settings.

GRAY (Number Please): Let me answer this by giving you my two favorite quotes: "Nothing drives basic science better than a good applied problem" (Newell & Card, 1985, p. 238) and "there is nothing so useful as a good theory" (Lewin, 1951). Basic research needs to keep in touch with real problems if it is to flourish. Applied research must be guided by theory or else it is just good old-fashion analytic thinking (rare but not owned by the cognitive community).

HOWELL (Decisions at Sea): Research is **good** if it conceived and conducted in a way that produces clearly interpretable results; **bad** if it isn't. The only difference is the scope of the questions posed and the breadth of the area over which the results generalize, and this is a continuum. Any "break point" is arbitrary.

SALAS (Decisions at Sea): No difference: The question we are trying to answer is a "what" question, not a "where" question (i.e., where the work is done).

STASZEWSKI (Harnessing Landmine Expertise): Stokes (1997) has made a compelling argument that this dichotomy can be a false one and that using a single bipolar dimension anchored by "basic" and "applied" to characterize all research is incomplete, misleading, and potentially detrimental to developing the theory, principles, and methods on which engineering depends, or, at least, should. His concept of "use-inspired" research is one that better fits the projects described here; they involved features of both basic science and engineering. No first principles or any theory existed on either the cognitive or behavioral aspects of skilled landmine detection; a fundamental understanding of expert performance and the processes that generated it had to be built before any instructional engineering could begin. Even then, an information-processing model of expert skill is not a training program; there still remained translating the models of expertise into a coherent and reasonably complete set of instructional activities. So, in short, these projects encompassed features traditionally associated with basic and applied research, and thus fit into the research space Stokes (1997) labeled "Pasteur's Quadrant."

HELMREICH (Get This ... on the Ground): Nothing is so practical as good theory-based research—Kurt Lewin.

DREWS (You Guys Better Take Good Care of Me): It is a problem of complexity. Basic research can afford the luxury of being reductionistic. This made it so successful in the past, but in being reductionistic implies that some of the work is irrelevant for applied contexts. The danger of applied work is that it gets lost in complexity, and, consequentially, no simple and unconditional explanations are possible.

Do you view your story as a success story? Why or Why not?

KLINGER: Yes. The customer was happy, and, relatedly, we helped them dramatically reduce their costs by improving performance. But I also think this was a success because several different HF disciplines converged. Team cognitive task analysis, individual CTA, organizational design, information management, and interface design were all applied to the solution set. This convergence provided for a rich solution. Also, the drills provided us with an opportunity to evaluate our impact.

BALL: Yes, because older adults are benefiting from cognitive training. The story is still in progress, however, as we try to determine how best to disseminate the training more widely.

GRAY: Definitely, yes. It was a success story in that we succeeded in stopping the purchase of an expensive system that did not deliver the functionality that it promised. It was also a success in that we were able to collect hard data on operational costs of the current versus proposed system and show that science was backed up by finance. It was also a success in that we showed that understanding multimodal, interactive behavior required us to analyze behavior at the level of elementary cognitive, perceptual, and action operations.

HOWELL: I guess so, from what I understand (although I have to rely on your account for most of the evidence of the operational changes that have occurred). I do believe there have been changes in training, but I don't know the extent of the design changes.

SALAS: Yes, by all means, absolutely ... practically, politically, personally ... by all means.

STASZEWSKI: Yes and no. Yes, because landmine detection training has been improved. No, because the work is unfinished; more remains that could and should be done to improve training for operators of hand-held mine detection equipment, at least until robotic systems are developed that can reliably detect landmines and remove soldiers from hazardous areas.

HELMREICH: Yes, the concepts developed changed the way first commercial airline pilots and then military pilots operate normally.

DREWS: I perceive the story as a success story because the integration of the existing literature will provide the ground for technological advancement. Also, the drug display in the OR has the potential to significantly improve patient safety.

What makes your story a "cognitive engineering" success story?

KLINGER: The short time frame, the customer satisfaction, and the convergence of the human factors fields mentioned above.

GRAY: My definition of *cognitive engineering* is "cognitive science theory applied to human factors practice." This is what we did. Theory led the way to success.

HOWELL: Well, since I have never been quite sure what the domain boundaries are that define it, my best guess is that the answer lies in the fact that a lot of the work was directed at determining how people and teams actually process information, interpret somewhat ambiguous and dynamic collections of diagnostic input, and do things in response. Come to think of it, I recall a lot of us exploring many of the same questions back in the 1950s to 1970s, but, not knowing any better, we called it "human information processing," "behavioral decision making," and, more broadly, "Human Performance" or "Engineering psychology." My point: I think a lot of effort is wasted trying to classify, distinguish, and label research that is fundamentally not all that distinguishable, and, in the process, all we wind up doing is confusing people—including important people in the "real world."

SALAS: This was one of the propellers that initiated Cognitive Engineering—one of a few that had embedded in it Naturalistic Decision Making (NDM) and Cognitive Engineering (CE), cognition in the wild, teams, expertise, training, and design with a balance of the best science applied to solve practical problems.

STASZEWSKI: Beyond cognitive engineering, the work represents a success story for the field of psychology, in general, and the discipline of cognitive psychology, in particular. This is because this work has linked basic research on human expertise to progress made against a difficult and significant problem. The projects' outcomes show that the foundation of scientific knowledge that these disciplines have accumulated via public investments in basic research has produced the kind of return on investment that those within and outside of these fields should find satisfying.

HELMREICH: It added the components of decision making, leadership, and situational awareness to the stick-and-rudder aspects of flight operations and flight training.

DREWS: The fact that the work applies a cognitive engineering approach to improve performance of clinicians and that there is clear empirical evidence for improvement.

What about this project was most intriguing?

KLINGER: For me, the most intriguing part of the project was trying to understand "what is really going on?" By taking a decision-centered approach, we were able to determine what this organization was trying to do. We kept asking them, "why is this team here, what is your mission, what do you as a group contribute to the overall mitigation of an event?" By asking these questions, we were able to pinpoint specific decisions that MUST be supported. From there, we could make a wide variety of recommendations (layout of the room, information channels, organizational changes, etc.) to support their decisions. It was also interesting to pose those types of questions. To ask an organization "why are you here in the first place?" provides interesting insight into what people think about their role, the role of their organization, and the perceived impact of both on the overall process.

BALL: Trying to understand why some older adults were having difficulties with everyday activities. In other words, what underlying cognitive abilities were predictive of each of the everyday tasks.

GRAY: Most everything. Saving seconds really does have a financial impact if you are the phone company. The fact that small differences in time when propagated over 50 million users led up to big bucks is pretty neat.

HOWELL: The fact that it represented the rare situation in which behavioral science was looked to for a "fix" to a high-profile, technology-related problem, AND WAS ALLOCATED THE TIME AND RESOURCES—**UP FRONT**—TO GIVE IT A FAIR SHOT. I think most of us involved saw it as a once-in-a-lifetime "put-up-or-shut-up" opportunity.

SALAS: How to get attention/credibility from customer/fleet; had to make it compelling.

STASZEWSKI: Two outcomes common to both of these projects were intriguing. The first was that relatively simple information-processing models could explain the initially daunting complexity of the behavior of the experts whose skills were analyzed. The second was that the training programs produced such unexpectedly large gains with so little instruction and task practice. The initial test results for project seemed too good to be true; it took successful replications to dispel my own skepticism.

HELMREICH: Trying to understand the full context of flight operations.

DREWS: Its complexity.

There is considerable debate today about how cognition should be approached in complex tasks? What approach to understanding "thinking at work" do you think will bear the most fruit?

BALL: Try to understand the task first. Obtain feedback from people performing the tasks, especially if they are experts. What types of situations make the complex task more difficult? What difficulties do novices experience in learning the complex task?

HOWELL: It's like the energy crisis and our dependence on oil—there are a number of ways to approach it; none represents a magic bullet; each can make a contribution; so we need to pursue them all (and hope for some productive convergence).

SALAS: Methods is the Achilles heel; multitrait-multimethod approach; measurement will bear the most fruit.

STASZEWSKI: Pressed to forecast, I suspect that the results of future efforts will show that some approaches and methods are better suited to some job environments and the tasks performed in them than others. Considering the fundamentally adaptive nature of expertise and the diversity of task environments, methodological eclecticism may well prove to be the most effective approach. The implication is the more diverse the methodological background and skills that cognitive engineers bring to a problem, the greater the chances of a successful outcome.

DREWS: For me the most promising approach is to use simulation. I think that high-fidelity simulation allows us to provide the context in which "cognition in the wild" can be studied without creating simplistic tasks that lead to artifacts. But it also allows us to create a situation where important parameters can be controlled. So, in my opinion, this has the potential to solve the old tension between ecologic validity and control.

In your opinion, what is the biggest barrier to a cognitive engineering success story (i.e., what is a barrier to achieving implementation in the field with measurable success)?

KLINGER: One of the biggest barriers is that we don't always know what we are going to do until we get inside of the domain to see where the problems and/or where the solutions lie. Linked to this is the lack of measures of effectiveness at the outset of many cognitive engineering efforts. We can't always say, "you will be this much more effective because we'll do this and that."

BALL: In my particular area, the process of implementing any change in the field is extremely slow. Changes are not made quickly or easily in public policy or health care, and thus it is extremely important to find "industry" partners early in the process, such that they can have input in the research phase.

GRAY: Two answers for you here. One for my initial reading of this question and one for my current reading.

Current reading answer. Ernestine was a success because a field trial had already been planned, and when it came out wrong we were the only ones that could explain the counterintuitive results (old technology faster than new technology). Most cognitive engineering is done in the background, and its results are never measured.

Initial reading answer. Cognitive engineering is hard to do. Currently, it requires intense training as a research cognitive scientist, combined with enough computer science and mathematics to be dangerous, but not enough to make a living as a computer scientist. It then entails being able to think about problems from the perspective of cognitive theory. This requires a coherent "worldview," such as is provided by unified architectures of cognition such as ACT-R, EPIC, and Soar. Today it requires the ability to build many of your own tools. It also requires being able to recognize when you are at the end of your science and either fake it or go into basic research to find a reasonably good answer. Most people are not all that good at all of these things.

SALAS: People do not know what we do. We are not good at marketing.

STASZEWSKI: The inability or unwillingness of some in our field to teach people outside of our field to understand what the disciplines of cognitive psychology and cognitive engineering are and do may be the biggest barrier to success. A related concern is the anti-scientific stance that some people in cognitive engineering have adopted. Engineering shouldn't be promoted or communicated as either mysterious or magical. Also, hand waving, obfuscation, or claims that "it's complicated" in response to questions posed by those without training in cognition will undermine successes that cognitive engineering has achieved and impede its development.

There are plenty of systems designers and engineers who believe their intuitions about human behavior and thinking are every bit as valid and useful as those of cognitive engineers. Our field must make a concerted effort to communicate and demonstrate that cognitive engineering applies a scientific knowledge base, not personal, idiosyncratic introspection and intuition.

DREWS: Complexity is the main problem. Especially, to develop an approach that allows one to measure the impact, but also still is sufficiently complex that it reflects the context of the application.

Becoming a Cognitive Engineer

In some ways, having succeeded implies that at some point in your past you had the right set of experiences. Beyond this, however, notice that our scholars point to the value of collaborators and mentors. Some found more value outside the classroom than in, but those who noted valuable formal training also noted experience with the real world. Just as the stories were different, so, too, are the skills that are brought to bear. Some of the researchers seem to advocate a more microscopic perspective that requires one to build in order to understand the complexity. Others advocate embracing the complexity directly. Finally, this section ends with some valuable, thoughtful advice from each of our scientists to the students interested in cognitive engineering.

Did your education and professional experience prepare you for this project? Why or why not?

KLINGER: Experience more than education. This was a large problem-solving exercise. We had to dissect it into small, workable pieces. From there, we were able to identify the critical elements and make recommendations. Previous work experience with Navy Sonar teams, Air Force weapons director teams, and Army Command and Control Centers helped us to know where to start and how to dive into the problem.

BALL: My education prepared me to a certain extent, but it did not prepare me for the difficulties and challenges of doing translational research in everyday settings.

GRAY: I had been moving into the new field of cognitive engineering since Don Norman coined the term in the early 1980s. Likewise, I had been following the exciting developments in cognitive task analysis since the LRDC group at University of Pittsburgh started doing their thing in the early 1980s. The GOMS work of Card, Moran, and Newell was something that I finally confronted and mastered in the mid-1980s. Having strong and knowledgeable colleagues such as Bonnie John was vital.

HOWELL: I sure hope so! After nearly a half century in the field, one would expect even a plant to have learned SOMETHING.

SALAS: No, schools do not prepare you for this. Grad school prepares you for a perfect world. You must learn adaptability and credibility.

STASZEWSKI: Past experience in performing information-processing analyses of expertise and building theories to explain how extraordinary human performance was achieved prepared me well for undertaking these projects. The theoretical and methodological orientation developed by working with Herb Simon and in the Carnegie-Mellon Psychology Department was extraordinarily beneficial.

HELMREICH: Yes, my mentor, Irving Janis, had great experience dealing with the real world as opposed to academic laboratory situations, in particular his study of the effects of bombing on civilian populations in WW2.

DREWS: My education prepared me for this project because it was very broad and always had both the basic and applied foci represented.

What training or skills of yours were critical to your story related in this volume?

KLINGER: I think that I had seen so many teams in action that I was able to look at the process rather than the outcome. In this case, we were here to improve the process. Certainly, a better process is more likely to result in a positive outcome.

BALL: I would not have been able to have the long-term program of research needed to answer my research questions without the ability to successfully compete for research funding. Thus, writing skills and the ability to convey the importance of this problem were critical. Training in experimental design and statistical analysis was particularly helpful.

GRAY: Skills? I had done a very deep-level cognitive task analysis of protocols collected from Lisp programmers back when I was interested in the Psychology of Programmers. This provided the right mindset and background experiences. I suppose anyone who has done a rigorous verbal protocol analysis of a complex task would also have the right skill set here. At least this would help. Unfortunately, I don't know that these skills are systematically taught anyplace. These are all research skills, so you would need to be a researcher or learn to think like one. Academic departments still do not teach task analysis of any form on a regular basis. Most of what is taught is really at too high a level to be of use in a true cognitive task analysis.

HOWELL: I'd like to think there were a number of them—ranging from the technical (my background in "human information processing and decision" research) to the interpersonal (leading a diverse group of Technical Advisory Board (TAB) members in arriving at consensus evaluative and directive feedback at meeting after meeting) to the communication ones (giving verbal and written feedback and writing reports). Of these, the latter two were probably most critical because there were other TAB members who were at least as well informed as I was on the technical aspects.

SALAS: I had a sense that there was a need to balance between science and practice. As a manager, I tried to maintain this balance.

STASZEWSKI: Several features that are characteristic of the "CMU approach" to analyzing skills that I have come to take for granted contributed substantially to the success of these projects. One is that whole, complete tasks are analyzed, not just parts (see Newell, 1973, for a discussion). Second, behavior is analyzed at a very fine-grain size, on the order of fractions of a second at a time, although the events being analyzed, of which there were many, often extended for minutes. Third, the idea that an explanation of expert performance involves a description of the processes that produce it influenced how the recorded observations were analyzed and findings synthesized. It also led naturally and directly to testing novices on the training programs described once the analyses of expert performance were completed. The test results not only measured the effectiveness of the instruction, they also served to validate the expert models, at least, to a degree. Fourth, the constructs of goal structures and productions heavily influenced how the experts' activities were parsed into events and were later related. Another important and, in this case, influential tenet of the CMU approach is that understanding how people perceive their tasks and learn to achieve their goals depends heavily on understanding the task environment—the information it provides, its structure, and the constraints it imposes. Finally, the achievements of Herb Simon, Allen Newell, and many of my CMU colleagues in analyzing and modeling problem solving, skilled performance, and learning instilled the assumption and confidence that explaining what, at first, seemed wildly complex and confusing was very tractable.

HELMREICH: Real-world experience in stressful environments gained during service as a line officer in the U.S. Navy.

DREWS: Being able to work on an interdisciplinary team and present theories and approaches of cognitive psychology and cognitive engineering to members of the team without such background. Being able to convince them that there is significant benefit to choose these approaches.

What advice do you have for young people interested in becoming cognitive engineers?

KLINGER: Be careful!!!! Above all, always be curious. Ask lots of questions and try to get to the center of the decision space. What decisions do people make and how can we support them? How to develop data-collection plans. It is difficult to know how and where to start when attempting to identify user requirements.

BALL: You will need persistence and patience. Do not give up if you are not immediately successful.

GRAY: I would worry about how they had heard the term. I would then advise them to read Don Norman's *Design of Everyday Things* and find a mentor that would lead them along those lines. I suppose I would also advise them to read my textbook, but that might take some time because I have not started writing it yet.

HOWELL: Figure out what a "cognitive engineer" is and then go for it. Personally I'd advise them to enroll in a strong cognitive psychology or cognitive science or "neck-up" type human factors program somewhere, but maybe that's just because I'm old-fashioned and have lived through too many fads.

SALAS: (1) Learn methods. Cook and Campbell's book on quasi-experimental design is a must. This is what we do in the field. (2) Be willing to get hands dirty, (3) Be flexible. (4) Have a passion for doing. (5) Believe in the science practitioner model.

STASZEWSKI: Learn as much about the existing body of scientific principles, theories, and methods of cognitive psychology as possible. Critically evaluate the knowledge and experience that have produced cognitive engineering success in past instances, and think carefully how these resources are relevant to the problem(s) at hand.

Two lessons from Herb Simon: (1) Select rich and challenging problems, and (2) "be sure to acquire as many good friends as possible who are as energetic, intelligent, and knowledge as they can be. Form partnerships with them whenever you can" (Simon, 1989).

No engineer can know too much basic science relevant to problem domain in which she or he has chosen to work. Successful cognitive engineering depends on a broad and deep understanding of the principles, theories, and methods of cognitive psychology.

I also believe that it is extremely important and helpful to be able to communicate what cognitive psychology and cognitive engineering are in a way the enables others outside of these areas to comprehend what we do and thereby understand how it might be able to help them. Being an effective teacher as well as a practitioner has great value.

HELMREICH: Learn the environment in which you are going to work, and don't presume that laboratory experience will either generalize or be credible to front line, real-world personnel.

DREWS: They should try to be as cognitively flexible as possible, and they should be playful and experimental in developing ideas.

Cognitive engineering in society

In this section, notice that many of the researchers found that interacting with the existing culture to be most challenging. These challenges were sometimes managerial, in getting the principal players on the same page, but many times the challenge was convincing the existing culture of the value of the new ideas. The scientists then reflect on how problems that involve convincing others might be solved. Finally, we gave our researchers a chance to offer suggestions to our policymakers. Some took us up on the offer.

What about your project was most challenging?

KLINGER*:* It was difficult to understand what was going on. At the outset, the room was full of people, many of them communicating via headsets. We had a difficult time determining exactly who was who, what they were doing, what they should be doing, and who they were passing information to. We needed individual interviews to help us dissect the intricate communication and decision channels within this complex organization. Also, over the years, several "work-arounds" were developed. By identifying many of those, we were able to piece together how the organization "should" work. However, identifying the work-arounds was a difficult task. It was hard to separate standard operating procedures and work-arounds.

BALL*:* Carrying out large, longitudinal studies to evaluate the impact of training. Trying to find or design new performance-based measures of everyday ability that could be used to evaluate transfer of training, rather than relying solely on self-reported difficulties.

GRAY*:* Getting the science right was challenging. The task posed by Project Ernestine required us to think much deeper about the fine-grained structuring of interactive behavior than we had thought before. In retrospect, we were on the leading edge of a change in how the cognitive science community thought about control of embodied cognition.

HOWELL*:* Getting the San Diego and Orlando folks to talk to each other (at either the bench or the management level) and at least pretending to be collaborating! Not a TAB meeting went by without this being Concern 1. Closely related was the challenge of getting the large herd of contractor cats headed in the same direction. To many of them, all that seemed to matter was their little "piece" and assurance that their funding would continue. We often found that they neither knew (or particularly cared) about what others were doing—even when the relevance for their work was obvious. For example, I recall two contractors developing entirely different measures for basically the same thing—and not even realizing it!

SALAS: Managing a large group of smart, well-intended people and to get them on the same boat/same mission.

STASZEWSKI: The most challenging and frustrating feature was overcoming skeptics of the PSS-12 training program who opposed its adoption. There were individuals who, even after criticisms were addressed, still dismissed the empirical results on the grounds that they "didn't believe them" and would not provide further arguments for their stance. Rational arguments based on solid empirical findings weren't persuasive in these cases, which was very frustrating. The only viable solution was to circumvent the opposition.

HELMREICH: Trying to change an embedded culture—and to overcome the suspicion of may pilots that psychology represented an attempt to brainwash them and their experience with programs that could most charitably be described as full of psychobabble. Dealing with managements that did not fully understand the interpersonal aspects of flight operations and wanted to do something on the cheap—can we do it with a 30-minute CD instead of a formal training program?

DREWS: To develop an understanding of the problem space of drug delivery anesthesiologists represent.

Some have bemoaned the difficulty of convincing sponsors that human factors, in general, or cognitive engineering, in particular, is an important consideration in all R&D efforts involving human operators. Do you think the difficulty is real or perceived? How did you overcome that barrier?

KLINGER: I think it is real. During my time at Klein Associates, I was surrounded by cognitive scientists, and we all understood the importance of our work. I am currently working at an information technology firm. It has been difficult to convince system and software engineers of the value of cognitive engineering and human factors. *Overcome barrier:* I'm still trying. Luckily, I have some case studies like the one reported here to fall back on.

BALL: This is a real difficulty. Within the transportation industry, vehicle manufacturers and highway engineers have only recently considered the physical, visual, and cognitive capacities of older operators when implementing new technologies and designs. *Overcome barrier*: By attending many meetings I might not otherwise have attended and presenting our research to a variety of audiences. Quite frequently I have observed that vision scientists are unaware of the work of psychologists or cognitive scientists and vice versa, and that both groups have information that could benefit the other one. Presenting at meetings for those actually involved in dealing with the public (e.g., Motor Vehicle Administrators, meetings, or insurance companies) has provided me with a new perspective.

GRAY: There are real difficulties, but they are not as large as the perception. The problem is to relate our science to problems that people want to have solved. It is generally accepted that the first half of the 20th century belonged to the physicists and the second half belonged to the biologists. The first half of the 21st century belongs to the cognitive scientists. The issues we confront today are cognitive science issues.

HOWELL: See next chapter, Bill Howell's commentary.

SALAS: Real. We don't market well, show how we matter, translate what we do. *Overcome barrier*: We were able to translate, showed motivation, showed value.

HELMREICH: Sometimes real, sometimes perceived.

STASZEWSKI: The problem is real. *Overcome barrier:* Considerable thought and effort were invested in responding to all questions about plans or actions by explaining as clearly as possible the science behind them and the reasoning involved in applying that science to the problem at hand.

DREWS: I think these are real difficulties. *Overcome barrier:* By being very persistent.

What advice do you have for our policymakers?

BALL: In my field, it is difficult to implement a change in public policy, at least in part, because the cost of implementing a change may occur in one agency, whereas the benefit (savings) may accrue to a different agency. Who bears the cost for improving cognitive function? Policymakers should try to see the big picture and have a method for balancing costs and benefits.

HOWELL: The only way to have a genuine impact on them, in my view, is to be prepared with solid input and a strategy for delivering it when crises or targets of opportunity arise. Three Mile Island, Vincennes, 9/11, Katrina, the medical errors report (or many less dramatic events like ill-advised budget proposals and legislative actions), having something defensible and relevant to offer, and a good strategy for delivering it can make a difference (bit by bit, piece by piece). But trying to snow them with brag sheets and past accomplishments is wasted effort. The fact that TADMUS may have brought about some positive changes is of absolutely no interest to them—unless or until there is another disaster. Right now, for example, Sen. Kay Bailey Hutchenson (R, TX) is leading a move to eliminate the National Science Foundation's Behavioral, Social, and Economic (SBE) directorate on all the usual grounds, despite countless hours spent by countless "advocacy" groups over the years (since the Proxmire "golden fleece" days) showing policymaker after policymaker how important this stuff is and what the investment has accomplished. If this move is defeated, it will not be because a lot of these folks have come to appreciate what we do. It will be because an effective campaign, focused narrowly on this one issue, is mounted by the collective science community. And next year, the whole issue could arise again.

SALAS: Have an open mind; listen; and, once you see we matter, support us.

STASZEWSKI: Both policymakers and industry managers responsible for developing high technology tools are more likely to achieve their goals if they ensure that human intelligence is included in the design and that the human and machine components are rigorously engineered.

HELMREICH: Listen to data.

DREWS: We have to be more aggressive in marketing our skills.

Suggested readings

Cook, T. D., & Campbell, D. T. (1979). *Quasi-experimentation: Design and analysis issues.* Chicago: Rand McNally College Publishing Company.

Hoffman, R. R., & Deffenbacher, K. A. (1993). An analysis of the relations of basic and applied science. *Ecological Psychology, 5,* 315–352.

Lewin, K. (1951). Psychological ecology. In D. Cartwright (Ed.), *Field theory in social science: Selected theoretical papers by Kurt Lewin* (pp. 170–187). New York: Harper.

Newell, A. (1973). You can't play 20 questions with nature and win: Projective comments on the papers of this symposium. In W. G. Chase (Ed.), *Visual information processing* (pp. 283–308). New York: Academic Press.

Newell, A., & Card, S. K. (1985). The prospects for psychological science in human–computer interaction. *Human Computer Interaction, 1,* 209–242.

Norman, D. A. (1986). Cognitive engineering. In D. A. Norman & S. W. Draper (Eds.), *User-centered system design* (pp. 31–61). Hillsdale, NJ: Lawrence Erlbaum Associates.

Norman, D.A. (2002). *The design of everyday things.* New York: Basic Books.

Stokes, D. E. (1997). Pasteur's Quadrant: Basic science and technological innovation. Washington, DC: Brookings Institution.

Chapter ten

Commentary by William C. Howell

Making a splash to make a difference: marketing human-centered design

As explained in somewhat more elegant terms in the Preface, the foregoing collection of success stories is basically a sales pitch. What's being pitched is a fairly new product called *cognitive engineering,* but the reality is that its most active ingredient—something called *human-centered design*—has been around for ages under brand names such as *Taylorism, human engineering, engineering psychology, ergonomics,* and *human factors.* Yet despite a rather impressive track record, very few ordinary citizens are aware of it, and even fewer fully appreciate its accomplishments or potential. That naturally raises marketing questions, several of which I examine rather closely in this chapter. In the interest of full disclosure, I must confess that I'm more interested in the generic ingredient than the cognitive engineering brand, and that bias is reflected throughout the chapter. Nevertheless, I think it is fair to say that all my observations apply equally to both.

In particular, then, I explore the reasons that human-centered design remains so stealthy and underappreciated, and I offer some observations on how the situation might be improved to benefit us all. Lest it get lost in the telling, the moral of this story is that technology is inherently fascinating, whereas its implications for us humans are not. Consequently, technology advances at warp speed whereas valid knowledge on how best to use it and what hazards it poses lags far behind—such risks going unrecognized until some high-profile disaster, such as those illustrated in the previous chapters, occurs, exposing the flaws. If this situation is to improve, those engaged in producing and applying human-centered design products (whom I refer to generically hereafter as *the human factors community*) must figure out better ways to market it and become more active in doing so. After analyzing some of the more prominent difficulties, including those inherent in the product, the producers, and past marketing efforts, I consider how success stories like the seven featured here can help improve the situation.

Before proceeding further, however, I should point out that not everyone agrees with my public-ignorance assertion. In a recent presidential address before the *Human Factors and Ergonomics Society* (HFES), for example, Wendy

Rogers presented an impressive array of Google-hit, popular citation, and other statistics suggesting that the bushel under which our collective candle had long been hidden has vanished, and the flame now shines brightly for all to see. Although I hope she's right, and I'm not going to contest her statistics, I've been looking hard in a lot of key places and have yet to see even a glimmer. So I'll continue with my story. If your experience is different from mine, feel free to stop here and go read some other piece of fiction.

Marketing problems inherent in the product

The commonsense illusion

As you're well aware, if you've read this far, the *human-centered design* concept embodies the simple philosophy that anything created for people to use (from living and working environments to complex electronic systems to simple hand tools) should be designed with human capabilities, limitations, tendencies, and preferences in mind. When they aren't, bad things can happen. However, careful consideration of these human characteristics at the design stage makes things safer and better, and, in general, the more precise the measures, the better the outcomes. Even if the user is neglected in the development process and an error-prone design evolves—a common occurrence, as these stories illustrate—the neglect can be rectified through modifications in either the thing (redesign), the human (training), or the combination of the two. All three approaches are represented in this book.

What makes this more than just common sense is that identifying and properly measuring relevant human characteristics is a daunting scientific undertaking, as is the process of human–machine system figuring out how to apply the information and assess the results. To be done right, there's nothing simple about any of this, as the stories so aptly illustrate. But the deceptive appearance of simplicity and obviousness has always made it difficult to sell people on the need for specialized research, tools, and experts. Consider, first, the design community.

Most engineers and other design professionals are preoccupied with how well the things they create (hardware, software, vehicles, weapon systems, etc.) work, paying far less attention to how well they suit the people obliged to use them. In the jargon of the field, designers tend to focus on the machine components, in what at some level is always a symbiotic human–machine system relationship. Over the past few decades, as the term *user-friendliness* has crept into the popular vernacular, the design community (reinforced by marketing pressures) has become increasingly sensitive to user acceptance. However, the idea that it takes more than common sense, the designer's personal experience, or some quick-and-dirty usability test to ensure that a new system will run smoothly and safely still has many skeptics. When tragedy does strike and human error is implicated, the tendency is still to seek out and punish the culprit, rather than look for human–machine system design flaws.

Fortunately, despite such skepticism, human-centered design has edged its way into a number of systems' applications, as illustrated in the previous chapters. Enlightened segments of the military, telecommunications, and aviation communities, for instance, have been drawing on it for over half a century and remain among its leading research sponsors. Subsequently, it has infiltrated a number of other domains, health care being among the most recent. Still most of these advances weren't initially welcomed with open arms by their respective design communities, and they often came about only in response to litigation or government pressure. Many design professionals still believe that it all boils, down to common sense and see little need for specialized research, tools, and experts.

For their part, users and the general public have been even less impressed—for exactly the same reason. They're unaware of how the nifty features they like came about and, if brought to their attention, would regard most as blatantly obvious. If they run across a feature they don't like, they can't understand why an obvious alternative didn't occur to the manufacturer. Certainly, in their view, it doesn't require some self-proclaimed human factors specialist or costly research effort to fix these things. They fail to realize, of course, that what they see as the obvious fix may well come with a host of less visible risks that only an informed study would reveal, rendering the cure worse than the disease.

The point, then, is simply that it is difficult to market as legitimate a field of endeavor rooted in a concept that most see as nothing more than glorified common sense. It is even more problematic when contributions from that field are difficult to trace and not widely recognized—our next marketing challenge.

The tracing and attribution problem

Most of the valuable human-oriented features of things we use didn't spring up overnight like mushrooms, but are the end product of painstaking cultivation extending far back in history. When they are recognized at all, as in the case of the present "success stories," only the last step in the process is noticed. This, of course, is not unique to human factors—few people know or care much about the fundamental research that was essential to the creation of the Salk vaccine, the cloning of Dolly, or space travel. So-called *applied research,* in which studies are designed explicitly to solve an identified problem, naturally gets more attention than the more open-ended basic kind.

In the case of human factors, however, the difficulty is exacerbated by the fact that the foundation underlying most successful applications is not easily identified and often lies in an entirely different field (or fields), like psychology or biomechanics. Human cognition (how people think), for example, has been the dominant focus of psychological research ever since a revolutionary shift from behaviorism occurred in the 1970s. Despite its impressive list of new tools and accomplishments, today's fledging cognitive engineering specialty would be a pretty hollow enterprise without the knowledge base accumulated by psychologists and other cognitive scientists over those 30-plus years.

Closer to home, consider the *TADMUS* story (chap. 8). The Navy did (under some pressure from Congress) invest heavily in what turned out to be a successful program aimed a fixing the human–systems failures involved in the catastrophic Vincennes incident. Yet what the story doesn't mention is that, for decades prior to this disaster, the Office of Naval Research (ONR) was among the world's leading sponsors of fundamental research on human decision making (as well as other cognitive functions). So when the SOS was sounded following the Vincennes disaster, the rescue effort wasn't obliged to start from scratch. There were plenty of well-equipped experts around, along with a well-stocked cache of tools and knowledge. Although the program ultimately opened up some new avenues in decision aiding and training (and repaved others), the odds that an effort of even TADMUS proportions could have accomplished what it did without this foundation are long indeed. Nevertheless, it would be extremely difficult, if not impossible, to trace exactly how each piece of background research figured into the end result.

The point here is that, although it is possible (if not easy) to find compelling stories in which human factors played a demonstrable role, such as the foregoing seven, they can never convey the full extent of the contribution because they're limited to immediate applications. In the vast majority of cases where human factors made a difference, there's no compelling story to tell because the most important contributions occurred much earlier in the process, and the end result involved a lot of other factors. Taking credit for anything good that came of it, even if technically defensible, produces nothing, but yawns and skepticism.

The abstractness, lack of pizzazz, and credibility problem

Closely related to the apparent obviousness of human factors contributions and the difficulty in isolating them is their inherent lack of "pizzazz." Very simply, most of our products don't have much inherent "sex appeal," and attempting to create some (e.g., by taking credit for high-profile accomplishments) poses a credibility risk. I was rudely awakened to this reality while serving as the chief scientist responsible for U.S. Air Force (USAF) research on human systems. Part of my job was selling military decision makers on the worth of our programs in order to sustain (and hopefully increase) the investment. What I discovered was that, no matter how conclusive my evidence, the best I could hope for was tolerance and grudging support. Brief them on a jazzy new weapon system or aircraft (or research toward those ends)—concrete things they can visualize—and their eyes light up; explain how a new method for training mechanics or detecting faults—abstractions—will improve force effectiveness, and the eyes glaze over.

What we have to offer, in other words, comes across as inherently abstract and mundane, so no matter how valuable or scientifically valid it may be, it generates little excitement. Moreover, if we come up with some finding that actually is surprising (which generally means contrary to personal experience), the inclination is to reject it and go with experience. Unlike physicists or chemists, what we're presenting are generalizations about humans, and because everyone is a member of that species, everyone features him or herself an expert. This is true for military decision makers, design professionals, the folks you see on Main Street, and, worst of all, our elected policymakers.

There's an ad currently running on national TV that, after identifying a number of common products the company does not manufacture, concludes by saying something like, "We don't make many of the things you use; we just make them better." This captures rather nicely the challenge in marketing human factors (broadly defined). It is a lot easier to impress people with concrete products (things they use) than it is abstractions (making those things better). If the abstractions seem obvious, trivial, or counterintuitive, the task just gets that much tougher.

The labeling and identity problem

As I mentioned at the outset, those practicing and promoting the human-centered design philosophy have done so under a number of banners, each of which has had special meaning for its particular sect, but little for anyone else. Psychologists and engineers who came together during WWII identified their newly forged collaboration as human engineering, but the engineering establishment objected, and this label was replaced in some quarters by human factors and in others by engineering psychology. Despite considerable overlap in membership and philosophy, each developed its own institutions and distinctive self-image. Meanwhile, elsewhere throughout the industrialized world, something called *ergonomics*—also heavily influenced by WWII—was making noises. To some, it was just *human factors* with an accent; to others, it was more about designing for human physical characteristics, whereas human factors included (perhaps even focused on) the mental (cognitive) ones. More recently, a number of other specialties have emerged, all fighting for independent recognition (e.g., human–computer interaction [HCI], computer-supported collaborative work [CSCW], macroergonomics, and cognitive engineering). Many in these latter groups consider human factors a relic from the "knobs-and-dials" era, whereas those who continue to identify with it (and after a long, bitter struggle now brand it human factors and ergononics [HF/E]) naturally take exception to that characterization. And the beat goes on.

I'm not suggesting that these and other specialties are just different labels for the same thing; they aren't—there are legitimate differences. I'm simply pointing out the marketing implications. Multiple brands with subtle distinctions only serve to confuse lay audiences. It's hard to market a concept like *human-centered design*, which has no commonly accepted name, image, or unified constituency.

Problems inherent in human factors professionals (broadly defined)

Not only does the nature of our product present marketing difficulties, so too do certain characteristics of the folks on the production and distribution end. I consider three: values, abilities, and strategies.

The dilemma of values

Throughout history—from the Garden of Eden to the invasion of Iraq—anecdotes and stories have been instrumental in shaping human events. Whether factual or fictional, prompted by noble or nefarious motives, we find a compelling story difficult to resist—even in the face of overwhelming evidence to the contrary. Attorneys, evangelists, merchants, journalists, terrorists, con-artists, and politicians have always understood this and have had no qualms in using it to their advantage. For scientists, in contrast, it poses something of a dilemma. Science prides itself on its objectivity and reliance on empirical evidence in drawing conclusions—in marked contrast to the selective use of stories and anecdotes to sell soap, presidential candidates, or wars. The mission of science is discovery, not persuasion. Moreover, until fairly recently, scientists have been reluctant to tout their findings in public venues for fear of misinterpretation or misuse, recognizing also that there is no payoff for taking such risks. Reputations and careers were built or destroyed largely on the basis of what their peers thought, not the public or elected officials. Although, as we see later, this picture has changed a bit, with most scientists now recognizing that promotion in some form is inevitable, many still don't like the idea and look to others (e.g., professional organizations, science reporters, federal funding agencies) to carry the ball on their behalf. There's probably no completely satisfactory way to resolve this dilemma, but the more tenaciously human factors researchers cling to the belief that hard evidence sells itself, and catering to public tastes is nonprofessional, the more difficult the marketing challenge becomes.

It should be recognized, of course, that not all those marching in the human-centered design parade identify themselves as scientists or are so viewed by others. A substantial number would describe themselves as practitioners, although even they recognize that their professional survival is ultimately tied to past and future research. So although a few outliers may feel comfortable matching tall tales with hucksters, lawyers, and politicians, most would share the scientist's inclination toward restraint.

The ability factor

The laudable reluctance to compromise values is not this community's only marketing handicap. With notable exceptions, scientists and human factors professionals are not known for their prowess in popular communication. Most of us have acquired at least passable proficiency in technical writing and can give a PowerPoint presentation without too much embarrassment. However, when it comes to writing an article for a popular magazine, attracting a capacity crowd to a public lecture, or convincing a congressman that our research is a national treasure, most of us, frankly, are pretty inept. We aren't trained for this kind of communication, and our scientific background forces us to attach so many caveats to anything even mildly interesting we might have to say that our story falls flat.

Strategic difficulties

Not only do most of us lack the ability to crank out spell-binding prose, we don't have a firm grasp of the institutions through which it is disseminated (notably the media and the public policy machinery). We aren't trained in this either, and without knowing the ropes, our chances of getting the exposure or reactions we want are minimal at best. For example, I've known distinguished researchers who, after being convinced that writing newspaper op-ed pieces is important, are shocked and dismayed when their painstakingly crafted masterpieces are rudely ignored. Failing to appreciate the volume of submissions that newspaper editors must act on daily, or the impossibility of providing feedback the way journal editors do, they take offense and vow never again to repeat *that* mistake, instead of persisting and eventually being heard. I have also accompanied professional colleagues on visits to congressional offices and seen them completely blow this valuable marketing opportunity by babbling incoherently, becoming confrontational, or coming in with no clear idea of what they wanted to communicate or its relationship to that office's agenda.

In summary, influencing public perceptions and policy requires a lot more than just technical knowledge, the ability to communicate with colleagues, and a good message. It requires the ability to communicate effectively on many levels and an understanding of both the audience and venue. This, in turn, requires specific training, strategic planning, persistence, and a high tolerance for failure on the part of those wishing to participate. Well aware of these requirements, most of the major scientific and professional organizations are taking the lead in coordinating public information and government relations activities for their constituents, including the compilation of success stories, the recruitment and training of willing participants, and strategic planning. Human-centered design issues are among those engaged by several such organizations. Viewed as a whole, however, the human factors community still has a way to go in preparing and organizing its troops to do battle effectively in the arena of public opinion, and qualified volunteers aren't breaking down the doors to sign up.

Why market human-centered design?

To this point, we have simply assumed that what we have to offer justifies a level of public support that it can never realize without raising its profile, and the way to do that is through more effective marketing. Accomplishment alone can't do the job, and we've considered some of the reasons why. However, before exploring how this relates to the foregoing success stories, I pause briefly to ponder the validity of these assumptions.

The human–technology support gap

Without realizing it or thinking much about the consequences, society has allowed—indeed encouraged—technology to forge ahead at its own frantic pace, leaving in its wake a host of unresolved human issues and unanticipated societal problems. Only now, for example, are we beginning to see the pot-holes and occasional road-side bombs scattered along the information superhighway. The reality is that no matter how ingenious, powerful, and truly amazing technological innovation may be, there is ultimately a point at which it must engage humans (in technical jargon, the *human–machine interface*). However sophisticated the automation, humans will always have the ultimate responsibility for landing the plane, administering the anesthetic, or firing the missile. And technology isn't infallible. Yet for all the reasons discussed, investment in the machine side has dwarfed that in the human side. More aggressive marketing, therefore, makes sense not just for advancing a science and profession, but because of the promise it holds for filling a lot of those nasty pot-holes.

The sea change in public research support

For most of its history, the nation's scientific enterprise was—by common consensus—largely self-governed, so the only master researchers needed to worry about was their collective body of peers. The American public and its elected officials didn't pretend to understand what science was up to, but were sufficiently impressed by the results ("miracle" drugs, space travel, computer technology, world military dominance, etc.) and were content to leave direction and management largely in the hands of the scientists. Thus, researchers could do their thing, and if they did it to the satisfaction of their peers and technically savvy program managers, they were generally rewarded with accolades and continued support. They didn't really need to explain their work to the folks on Main Street or justify it to the folks on Capitol Hill. Applications played a bigger part in funding decisions at some agencies than at others, but the overarching emphasis was on supporting quality research in recognized disciplines (in the belief that good things would come of it), rather than directing science to engage and solve society's problems.

But over the past two decades, the nation's largely hands-off attitude regarding its research investment has undergone a sea change, with scientists finding themselves in an entirely new ball game—one to which they often come ill-prepared and regard as somewhat offensive. It is a game in which the unwashed masses, rather than just the informed few, have a loud voice in picking winners and losers. In this venue, one good story, convincingly told, can often trump a truckload of data. Furthermore, it is a game in which the traditional scientific disciplines have become largely irrelevant, replaced by society's fixation on its many problems (and its misguided belief that science can resolve them all). Politicians who once were applauded (and reelected) by constituents for underwriting advances in chemistry, biology, physics, and even psychology now must justify public investment in terms of curing diseases, reducing the terrorism threat, or improving K-12 education. To survive and succeed—or have any impact at all on society—scientists must compete for attention in the increasingly crowded and noisy marketplace of self-promotion.

So if our assumptions about the societal value of human-centered design and its currently low profile are valid, it follows that in today's political environment, the only way to make a difference is to make a splash. Moreover, the shift in research support from a disciplinary to a practical-applications orientation carries two strategic implications. First, if our goal is to market a science-based approach to solving important practical problems, doing so by distinguishing and promoting disciplinary brands like HF/E or cognitive engineering is misguided and probably counterproductive. Second, success in this new environment must start with recruiting, reprogramming, and training a lot of professionals, and generating persuasive promotional material along the lines of the present success stories. Once again, there are other reasons for creating and promoting distinctive brands, some probably more legitimate than others, but that's of no consequence here. This chapter is concerned exclusively with marketing human factors, broadly defined, not serving all the other needs of the professionals involved.

So what are the prospects for making a splash, and how do these stories fit in?

In view of all the challenges involved and the nature of today's public support landscape, can human-centered design be marketed more effectively? The answer, in my opinion, is that, although difficult, it is far from impossible, and it starts with recognizing and addressing our limitations. Narrowing the gap between technology's advancement and its effective adaptation to human users is somewhat akin to addressing the world energy crisis: It is easy to understand but difficult to sell, and there's no "magic bullet."

To make headway in either calls for the collective application of a *variety* of resources and strategies, let us consider some, beginning with our seven success stories.

Expanding the content pool

These stories and the book overall provide a glimpse into a fascinating world that few outside the field have had occasion to visit. But it's just a glimpse. Unless non-technical readers are drawn to it in sufficient numbers and stimulated to look further, and on doing so encounter other fascinating material, its effect will be minimal. Some material written for lay audiences is available, but not nearly enough. For example, Steve Casey's and Don Norman's books have been well received, and the quarterly magazine, *Ergonomics in Design,* cranks out a steady stream of articles intended for public consumption, although its readership remains largely confined to the professional community. Of course there are Web sites and house organs. The point, however, is that our current supply of compelling promotional material is too small, too limited, and too haphazardly distributed to support any concerted marketing effort. Although it remains to be seen how many ordinary citizens this collection of success stories will reach, it constitutes a potentially powerful addition to our rather meager pool of material, and hopefully will stimulate others who have stories to tell to step forward, providing them excellent guidance for doing so.

Expanding and preparing the sales force

If making a splash where it counts is the goal, an expanded content pool is just one element. Knowing how, when, and where to jump in is equally important, as is the cultivation of an expanded sales force. At the most fundamental level, the entire human factors community needs to become more actively involved, savvy, and proficient in promoting its core product. This entails acquiring a more complete understanding of target audiences (for starters, the general public and policymakers) and the venues through which they are reached. It also requires an honest assessment of our individual and collective communication deficiencies and taking steps to address them. For example, instead of getting hung up on excruciating details in the telling of what, to us, are fascinating stories (as scientists are prone to do), we need to step into the intended listener's shoes. What would be of greatest interest to them, and how can the message and delivery be crafted to stoke their interest—even if doing so requires some trimming and shaping? Because of its negative connotations, I've avoided using the term *spin,* but that's basically what I'm suggesting here. Karl Rove and James Carville notwithstanding, spinning doesn't necessarily require bending the truth.

Not everyone, of course, can read audiences or master the tricky art of putting the right spin on a story without compromising its core truth, but I believe there are far more human factors professionals who could develop and employ such skills than are currently doing do. Sustained recruitment and education, therefore, is essential. Those completely devoid of aptitude in this regard will likely continue to self-select and stick close to their day jobs, as indeed they should. But for the rest, briefly revisiting our seven stories may provide some useful insight.

Putting stories to use

Each of these stories met four rigorous requirements set forth by Cooke and Durso; hence, legitimacy was ensured, and general appeal was encouraged. I leave to your judgment how intrinsically compelling they are. In the hands of an expanded sales force, however, the issue of relevance becomes paramount. Six application contexts are represented (military, highway transportation, air transportation, nuclear power, medical, and telecommunications), together with two principal intervention strategies (design and training). Obviously, it would be important to draw selectively from this collection in pitching a new training research program to some federal agency or persuading some project manager to get cognitive-engineering input in developing a new product.

For the media, operating as it does in real time, "newsworthiness" (typically in relation to current events) as well as intrinsic appeal is critical. Hence, if our salespersons are alert to opportunities and jump in immediately with something relevant, their chances of closing the deal increase. Recent publicity over the role of hand-held cell phones in auto accidents, for example, happened to coincide with a special section in the journal, *Human Factors,* showing that cell phones are but a part of the comprehensive driver distraction problem. Reacting quickly with a press release, the Human Factors and Ergonomics Society staff was successful in getting this important message to a national audience and relevant congressional offices and, in the process, striking a blow for human-centered design.

Timing is also important for politicians, but what matters most to them is how the story (or delivered message) bears on issues currently on their radar or what direct implications it has for their constituency. The shelf life of any message delivered to a congressional office, no matter how powerful or persuasive, can be measured in minutes, if not directly related to one of these two priorities. For other audiences, of course, still other considerations apply. Here as elsewhere, the effectiveness of any success story, including the present ones, is as dependent on the particular mindset and readiness of the intended audience as it is its general interest value.

And that's probably as good a note as any on which to end this chapter. Good stories, employed strategically by an expanded sales force of savvy, well-prepared human factors professionals, *can*—with the help of professional organizations—make a splash and, in so doing, ultimately make a difference. The seven you see here represent a good start in that direction.

References

Casey, S. (1998). *Set phasers on stun: And other true tales of design, technology, and human error.* Santa Barbara CA: Aegean.

Casey, S. (2006). *The atomic chef.* Santa Barbara, CA: Aegean.

Ergonomics in design. Santa Monica, CA: Human Factors and Ergonomics Society.

Lauber, J. (1997). Changing the culture of safety: the aviation industry. In W. Rouse, N. Kober, & A. Mavor (Eds.), *The case for human factors in industry and government* Washington, DC: National Academy Press.

Norman, D.A. (2002). *The design of everyday things.* New York: Basic Books.

Norman, D. A. (2004). *Emotional design: Why we love (or hate) everyday things.* New York: Basic Books.

Rogers, W. (2005). Who knows about human factors and ergonomics? A lot more people than you might think. *Human Factors and Ergonomics Society Bulletin, 48,* 1–2.

See Winter 2004 issue (Vol. 45). *Human factors.* Santa Monica, CA: Human Factors and Ergonomics Society.

Index

A

B

C

D

E

F

G

H

I

J

K

L

M

T

U

V

W

X

Y